Andrology

Progress in Reproductive Biology

Vol. 3

Series Editor: P.O. HUBINONT, Brussels
Assistant Editor: M. L'HERMITE, Brussels

Editorial Board: E. BEAULIEU, Paris; E. DICZFALUSY, Stockholm; K. EIK-NES, Trondheim; J. FERIN, Louvain; P. FRANCHIMONT, Liège; F. HEFNAWI, Cairo; J. KELLER, Zürich; B. LUNENFELD, Tel-Aviv; L. MARTINI, Milano; J. MULNARD, Brussels; F. NEUMANN, Berlin; M. NEVES e CASTRO, Rio de Janeiro; H. PEDERSEN, Copenhagen; J.C. PORTER, Dallas, Tex.; P. PUJOL-AMAT, Barcelona; R.J. REITER, San Antonio, Tex.; P. SOUPART, Nashville, Tenn.; H.-D. TAUBERT, Frankfurt a/Main; M. THIERY, Gent

S. Karger · Basel · München · Paris · London · New York · Sydney

Andrology

Basic and Clinical Aspects of Male Reproduction and Infertility

Volume Editors
JERALD BAIN, Toronto, Ont.; E.S.E. HAFEZ, Detroit, Mich. and
B. NORMAN BARWIN, Ontario, Ont.

29 figures and 30 tables, 1978

S. Karger · Basel · München · Paris · London · New York · Sydney

Progress in Reproductive Biology

Vol. 1: Sperm Action. 5th International Seminar on Reproductive Physiology and Sexual Endocrinology, Brussels 1975. Eds.: P.O. HUBINONT; M. L'HERMITE, and J. SCHWERS, Brussels.
XII + 314 p., 89 fig., 55 tab., 1976. ISBN 3-8055-2244-4

Vol. 2: Clinical Reproductive Neuroendocrinology. 6th International Seminar on Reproductive Physiology and Sexual Endocrinology, Brussels 1976. Eds.: P.O. HUBINONT; M. L'HERMITE, and C. ROBYN, Brussels.
X + 286 p., 82 fig., 27 tab., 1977. ISBN 3-8055-2382-3

Cataloging in Publication
  Andrology: basic and clinical aspects of male reproduction and infertility
  volume editors, Jerald Bain, E. S. E. Hafez, and B. Norman Barwin. – Basel; New York: Karger, 1978.
  (Progress in reproductive biology; v. 3)
  Based on the 1976 Canadian Andrology Society meeting in Toronto.
  1. Reproduction – congresses 2. Sterility, Male – congresses
  I. Bain, Jerald II. Hafez, Elsayed Saad Eldin, 1922– ed. III. Barwin, B. Norman, ed. IV. Canadian Andrology Society V. Title VI. Series
  W1 PR681B v. 3 / WJ 709 A574 1976
  ISBN 3-8055-2807-8

All rights reserved.
No part of this publication may be translated into other languages, reproduced or utilized in any form or by any means, electronic or mechanical, including photocopying, recording, microcopying, or by any information storage and retrieval system, without permission in writing from the publisher.

© Copyright 1978 by S. Karger AG, 4011 Basel (Switzerland), Arnold-Böcklin-Strasse 25
Printed in Switzerland by Tanner & Bosshardt, Basel
ISBN 3-8055-2807-8

Contents

Foreword VI
Preface VII

HAFEZ, E.S.E. (Detroit, Mich.): Anatomical Parameters of Male Reproduction 1
PARRISH, R.F.; POLAKOSKI, K.L., and ZANEVELD, L.J.D. (St. Louis, Mo.): Standards of Analysis and Clinical Biochemistry of Semen   12
BAIN, J. (Toronto, Ont.): Neuroendocrine Parameters of Male Fertility and Infertility 33
DAHLBERG, B. (Malmö): Infection of the Male Reproductive Tract ...   46
ANSBACHER, R. (Houston, Tex.): Immunological Aspects of Infertility   60
SHOKEIR, M.H.K. (Saskatoon): Genetic Aspects of Male Infertility ...   70
KLOTZ, P.G. (Toronto, Ont.): The Urologist and Male Infertility ... 103
MISKIN, M. and BAIN, J. (Toronto, Ont.): The Use of Diagnostic Ultrasound in the Evaluation of Testicular Disorders 117
BARWIN, B.N.; MCKAY, D.; JOLLY, E.E., and HUDSON, R.W. (Ottawa): Pharmacological Therapy in Male Infertility.................. 131
BARWIN, B.N. (Ottawa): Artificial Insemination and Semen Preservation 141
BERGER, D.M. (Toronto, Ont.): Psychiatric and Psychological Aspects of Male Infertility 157
SHAUL, D.L. (Toronto, Ont.): Newer Concepts in Marital and Sex Therapy 165

Contributors 180

Foreword

There has long been a particular group of skeletons in the closets of reproductive medicine. These ossified enigmas are easily recognizable to any anthropologist, and just as surely forecast by the student of reproduction to be masculine. Why has the male been the skeleton in our closet? Again, the answers come easily. Early interest was in domestic animals, and females were more valued and useful as food and food producers. Artificial insemination has only recently become a scientific pursuit. Male potency has been a restricted clinical area in a world populated by male scientists and physicians. The nature of spermatogenesis is complex and the long delay between spermatogenesis and shedding in the ejaculate are frustrating to both student and practitioner. Add to these the recent advent and application of tools such as electron microscopy and radioimmunoassay. Finally, reflect upon the notoriously poor clinical results in dealing with male versus female infertility, and the totally depressing state of the art until about 10 years ago will be obvious. Recently, expanding technical skills, new descriptive knowledge and a real interest in the control of male fertility has brought the skeletons out of our closets. This book bears witness to the rehabilitation of studies in the basic and applied aspects of male reproduction – andrology. The 1976 Canadian Andrology Society Meeting in Toronto inspired the text. Since Canada has been a leader in this area, it is no mere happenstance that the bulk of the contents of this volume originated in Canadian institutions. The progress evidenced is long overdue and must be continued. Clearly, one of the most important messages to be found in these pages is that andrology has come of age and will not be a closet science in future. For this, the contributors, their colleagues and forebearers deserve our congratulations and support.   FREDERICK NAFTOLIN

Preface

Andrology has come a long way, rising from a nonexistent biological discipline to a well-recognized branch of the medical and life sciences in a relatively few short years. The study of male reproductive processes has expanded to the point where there are now enough andrologists to gather together at meetings and seminars. This book is the product of one such gathering, an andrological meeting in Toronto, Canada in 1976 which was the stimulus for a joint effort by the editors to compile works of merit and importance within the covers of one book.

This book has a distinct clinical function – intentionally so. Attempts to influence male reproduction in clinical medicine are increasing and as they increase so does our knowledge.

The editors are grateful to the authors for the effort they have exhibited, for their time, their patience and above all for the excellence of their contributions. The editors would also like to thank S. Karger AG for their patience in dealing with us and particularly Denise Greder who was so helpful and efficient. Finally, we are indebted to the secretarial assistance of Morag Smith who was responsible for all correspondence, for typing revisions and for manuscript compilation.

The Editors

Anatomical Parameters of Male Reproduction[1]

E.S.E. HAFEZ

Reproductive Physiology Laboratories and Andrology Research Unit, C.S. Mott Center for Human Growth and Development, Wayne State University School of Medicine, Detroit, Mich.

Progress in methodology of electron microscopy, analytical biochemistry and enzymology, and tissue culture have contributed to recent advances in male reproductive processes [8, 9]. The purpose of this chapter is to summarize modern concepts of functional anatomy in the male as they apply to clinical andrology.

Spermatogenic Functions

Several parameters have been used to evaluate the number of successive spermatogonial divisions during the seminiferous epithelial cycle: (a) the mitotic index of the spermatogonia after blockade with colchicine in which localization of the spermatogonial mitosis are localized during a seminiferous epithelium cycle; (b) the labeling index of thymidine-labeled cells to identify DNA synthesizing spermatogonia; (c) the labeled mitosis index at various intervals following injection to determine both DNA synthesis and spermatogonial mitosis durations; (d) the labeling intensity (number of silver grains per nucleus) of the spermatogonia one or several divisions after labeling to estimate the number of divisions undergone by one type of spermatogonia; (e) nuclear variations of spermatogonial populations, and (f) numerical ratio of different spermatogonia at various stages of cycle which estimate the number of spermatogonia divisions [3, 18].

The normal testis is characterized by certain features: large diameter of the tubules; a thin but definite basement membrane; a thin tubular wall two

---

[1] Supported by a grant from the Ford Foundation.

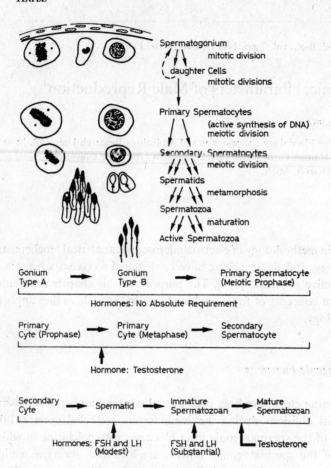

Fig. 1. Top: The stages through which a spermatogonium with the diploid number of chromosomes must pass in order to develop into an active spermatozoon carrying the haploid number and either an X or a Y chromosome. The active cell is represented to the left, the inactive to the right. All stages, with the exception of final maturation and activation of the spermatozoon, take place within the seminiferous tubule. *Bottom:* Hormonal control of spermatogenesis encompasses developmental stages from the prophase of meiosis through final maturation of spermatozoa. Testosterone is required throughout. LH stimulated testosterone production. FSH is synergistic with LH and in addition stimulates growth of the seminiferous tubule and Sertoli cell secretion [18].

or three layers thick; and full and regular spermatogenic activity from basal spermatogonium, primary spermatocyte, secondary spermatocyte and spermatid to terminal spermatozoa. The typical adult A pale or A dark spermatogonia only appear during puberty.

The lumen is often present especially if tubular section is exactly transverse, commonly containing sloughed spermatogenic cells. Only a few spermatozoa can be noted as once formed they are dislodged into the lumen. The initiation and maintenance of spermatogenic activity are under delicate endocrine control (fig. 1).

Spermatogenesis consists of a series of stepwise morphological changes which proceed in a prescribed schedule. This pattern produces a regular series of cycles of spermatogenic activity as the result of phases of cell division followed by maturation. The result is a series of characteristic combinations of cells resulting from the fact that the cell divisions are synchronous [10].

The seminiferous tubules are surrounded by myoid cells with contractile activity. The transport of immotile spermatozoa from the testis is facilitated by the contractile activity of myoid cells, the flow of the fluid, and the movement of kinocilia. The testicular capsule undergoes periodic contractions and relaxations. It seems that these rhythmic contractions and relaxations serve to massage the seminiferous tubules and facilitate the transport of nonmotile sperm from the seminiferous tubules toward the efferent ducts [4].

The intertubular tissue contains Leydig cells in loose connective tissue, fibroblasts, reticular cells, macrophages, plasma cells, lymphocytes and mast cells, and histocytes scattered among fibrils and occasional collagenous fibers. The connective tissue may reflect testicular function. Elastic fibers in the tubular wall appear at puberty, are absent in postpubertal hypogonadism, and appear precociously in precocious puberty, but are present in patients with arrested spermatogenesis [6]. The tubular wall may undergo hyalinization in various conditions associated with androgen deficiency. The intertubular connective tissue usually becomes more fibrous with regression of the tubules in advanced age, in testicular ischemia, and in several pathological conditions [12].

An effective blood-testis barrier separating seminiferous epithelium from the general circulation seems to be formed by special cells of the basement membrane of the tubule and by some special features of the Sertoli cells [7]. Although they are different anatomically, the blood-testis barrier serves in much the same fashion as the blood-brain barrier.

Sertoli Cells

Sertoli cells are triangular cells found next to the basement membrane of the seminiferous epithelium. The cells have an oval nucleus with folded nuclear membrane and large nucleolus. The cytoplasm has elaborate thin processes surrounding the adjacent germ cells except for the spermatogonial stem cells contacting the basement membrane. The basal portion of the Sertoli cell contains its characteristically lobulated nucleus with a prominent nucleolus. The cytoplasm contains slender spindle-shaped 'crystalloids of Charcot-Bottcher', the chemical nature and function of which are unknown [19].

Sertoli cells do not divide in the mature testis, and they are very resistant to ionizing radiations (X rays) and to various toxic agents which cause degeneration of the germinal epithelium.

Sertoli cells present polymorphous appearance in both their nucleus and cytoplasm since they undergo changes of shape and activity in relation to the seminiferous epithelium cycle. After the release of spermatozoa, the size of the Sertoli cell decreases.

With advancing puberty structural and ultrastructural changes take place in the germinal cells and Sertoli cells. With the appearance of the lumen in the seminiferous tubules the cytoplasm sends numerous lateral expansions between the neighboring germinal cells which also interdigitate with those of the Sertoli cells. Numerous intermediate forms (immature Sertoli cells) appear at the onset of puberty.

Sertoli cells perform several functions: (a) they play a role in the coordination of spermatogenesis and metabolic exchange of germ cells before their release in the lumen of seminiferous tubules; (b) they have considerable phagocytic activity as shown by the presence of lysosomes, and (c) they have an endocrine function.

The physiological significance of Sertoli cells in spermatogenesis is not fully known. They provide mechanical support for the developing germ cells and play a role in the release of the mature spermatozoa from the tubule. Sertoli cells are the obligatory anatomical pathway for the metabolic interchange between germinal cells and the blood stream. Sertoli cells also play a role in the resorption of the residual body of spermatozoa and resorption of degenerating spermatogonia [19].

Sertoli cells produce androgens, particularly testosterone and derivatives of pregnane and pregnene [16]. The residual bodies provide a substrate for the synthesis of these steroids. Lipid-rich bodies undergo phagocytosis by the Sertoli cells and are eliminated during the transformation of the spermatids into spermatozoa. Sertoli cells may be engaged in the production of hormones presumably of an estrogenic nature [15], and may act as an intermediate target system for gonadotropins.

Leydig Cells

Leydig cells are polyhedral, epitheliod cells with a single eccentrically located nucleus which is spherical or ovoid and distinctly vesicular with its three eccentric nucleoli. The cytoplasm of the mature Leydig cell is usually abundant, finely granular, and is stained by many acid dyes but has little affinity for the usual basic dyes. Leydig cells containing an abundance of large lipid droplets are likely to contribute relatively small amounts of testosterone to the circulation.

Leydig cells are scattered among the seminiferous tubules forming piles, cords or clusters around the sinusoidal capillaries with which they are closely related [13]. They are often irregularly grouped in the angular spaces or in strands along the intertubular spaces.

Leydig cells originate from fibroblasts during fetal life as well as prepubertal and postpubertal life. VILAR [22] noted several types of Leydig cells in men of different ages which vary in size, shape, and histochemical characteristics: types A, B, C, and D fibroblast-like cells, fetal and adult Leydig cells and degenerating cells. These represent progressive steps in the transformation from the juvenile fibroblast of the mesenchyme (fibroblast A) to the fetal or adult Leydig cells. At birth Leydig cells are present due to the effect of circulating maternal gonadotropins, and disappear a few weeks after birth. Thus in the prepubertal testis, the interstitium contains no Leydig cells. With the onset of puberty, only mature fibroblasts and Leydig cells persist, whereas degenerating forms decrease in number. In the postnatal life Leydig cells develop from mesenchymal or fibroblastic cells.

Scrotum

The testes are maintained in the scrotum at a temperature several degrees below body temperature by contraction or relaxation of the cremaster muscle. This lower temperature is necessary for normal spermatogenic activity. Contraction and relaxation of the dartos muscles raise and lower the testes with respect to the body, and their temperature is thereby regulated.

Also, the pampiniform plexus, which is formed by a network of veins as they leave the posterior border of the testes and pass up the spermatic cord, surrounds the convoluted testicular artery and serves as a thermoregulatory mechanism.

The descent of the testes involves abdominal migration to the internal inguinal ring, inguinal migration through the canal, and finally migration within the scrotum. Unilateral or bilateral cryptorchidism (failure of testicular descent) causes defective spermatogenesis and decreased androgen production, as a result of increased temperature. Elevation of testicular temperature as a result of cryptorchidism causes dramatic changes in testicular metabolic activity, e.g., decrease in glucose uptake, increase in metabolism of glucose to CO_2, and incorporation of lysine into protein [17]. In the ectopic testis steroid-3B-ol dehydrogenase in Leydig cells decreases [12].

Efferent Ductules (Ductuli efferenti)

The epithelial lining of the ductuli efferentes is made of kinociliated cells alternating with stereociliated cells and nonciliated cells whereas the entire ductus epididymidis is lined by pseudostratified columnar epithelium made of stereociliated cells and nonciliated cells. In the more distal portions of the ductuli efferentes, characterized by narrow lumen, nonciliated cells with signs of resorptive activity predominate. The nonciliated cells of the ductuli efferentes have both secretory and absorptive functions. The kinociliated cells of the ductuli efferenti are involved in the transport of spermatozoa from the testes to the epididymis. The fibromuscular capsule contains the same cellular and extracellular elements with abundant collagen deposition.

Epididymis (Ductule epididymidis)

Eight segments can be recognized in the epididymis as judged by the diameter of the lumen, the height and cytological characteristic of the epithelial cells, and cell organelles and cell surface [11]. The lumen is distended in the caput region, very narrow at the transition of the corpus and cauda. The narrow transition between the corpus and cauda is characterized by folds of the epithelium.

The epididymis is lined with a pseudostratified epithelium; small angular cells form a discontinuous basal layer in contact with the basal lamina, and taller columnar cells reach the free surface where they exhibit a tuft of stereocilia. Epithelial cells of the tail segment show absorbing and apocrine secretory activity.

The smooth muscle-like cells in the efferent ductules and epididymis exhibit spontaneous, rhythmic contractions, whereas the vas deferens shows no spontaneous contractions. The tail of the epididymis and the vas deferens share the same nerve supply as the ductus deferens.

The epididymis receives an innervation to the smooth muscle of the ductus epididymis commencing at the mid-corpus region. This innervation contains both NE- and AChE-positive fibers and is excitatory and possibly inhibitory to the smooth muscle.

The subepithelial layer of the excretory duct system consists of a thin musculature cell which increases in proximodistal direction. The ductuli efferentes and the initial segments of the ductus epididymidis have an inner circular musculature and an outer layer of longitudinally and obliquely arranged muscle bundles. The musculature contains three types of contractile cells. Smooth muscle cells of the cauda epididymidis and vas deferens are characterized by uniform size, close arrangement, even distribution and thin myofilaments with numerous dense patches. Presumably, different types of smooth muscle cells receive different types of adrenergic innervation.

The epididymis performs several functions. While transported from the testis to the ejaculatory duct through the epididymis, spermatozoa undergo maturation and achieve motility and fertilizability. Spermatozoa are stored in the cauda epididymidis and vas deferens. Depending on the frequency of ejaculation, there is an equilibrium between breakdown of aged spermatozoa and fullness of the epididymis. During their transport in the epididymis, spermatozoa are exposed to differing biochemical milieu with varying concentrations of phosphorylcholine, carnitine, glutamic acid, sialic acids, steroids and a variety of enzymes. During their maturation, epididymal spermatozoa acquire or get rid of varying amounts of sodium, potassium and bicarbonate. A few disulfide bonds are also added to both the chromatin and tailpiece. Maturational changes in epididymal spermatozoa involve the cytoplasmic droplet acquisition of independent motility, molecular characteristics of the surface of the plasmalemma, and structural cross-linking of protein-bound thiols in head and tail organelles.

Spermatozoa are transported from the testis and through the efferent ducts by pressure of rete testi fluids, aided by the beat of kinocilia and the peristaltic movements of the muscle wall. The Sertoli cells seem to contribute to the rete testis fluid which transports spermatozoa tubules into the epididymis.

Contraction of the smooth muscles of the epididymis stimulated by intraluminal pressure, action of catecholamines and neurohypophyseal hormones (oxytocin and vasopressin) caused by flow of sperm and fluid from the testis, and peristaltic movement of parts of the epididymal tract, have been implicated in the transport of spermatozoa [14].

With frequent ejaculation, progressive fatigue of the seminal emission reflex may contribute to low sperm counts. It is possible that sperm counts may increase after long periods of sexual abstinence.

The time of transit of spermatozoa through the epididymis varies from 12 to 21 days depending on the techniques employed for study [21]. About half of the spermatozoa which enter the head of the epididymis disintegrate before reaching the cauda region which stores mature spermatozoa. Although the epididymal environment is favorable for survival of spermatozoa, they are not preserved indefinitely in a viable state; they gradually age and lose their viability. After prolonged sexual rest, spermatozoa first lose their fertilizing ability, then their motility, and finally disintegrate in the ductus deferens, resulting in poor semen quality. Unless older spermatozoa are eliminated from the male tract at regular intervals, their incidence in the ejaculate increases with less frequent copulation. Thus, after long periods of sexual rest, the first few ejaculates contain more aged spermatozoa.

Vas deferens (Ductus deferens)

The luminal epithelium with its underlying basement membrane is surrounded with connective tissue made of collagen fibers. The vas deferens is characterized by the presence of a thick, well organized and richly innervated smooth muscle coat. The proportion of longitudinal to circular muscle fibers as well as the overall thickness vary from the epididymal to the urethral end of the vas deferens. The vas deferens and most of the accessory genital glands have a very dense supply of adrenergic terminal which is resistant to hypogastric denervation as well as to lumbosacral sympathectomy. In comparison to the proximal portion of the epididymis where adrenergic terminals are sparsely distributed and principally found around blood vessels, the cauda epididymis and vas deferens exhibit rich networks of sympathetic fibers terminating directly at the musculature [5]. The terminals which contain norepinephrine are associated with short postganglionic adrenergic neurons.

Lack of testosterone during postnatal growth of the reproductive system causes the reduction in the norepinephrine content of the vas deferens before and after sexual maturation [2]. It would appear that male sex hormones influence the development and maturation of the short adrenergic neurons, probably via an action on the hypothalamus [1]. The vas deferens transports spermatozoa from the tail of the epididymis to the urethra by peristaltic movements which may occur during precoital stimulation and courtship.

Accessory Sexual Glands

The accessory sexual glands are organized as tubular or alveolar structures whose functions are either secretory or absorptive. The ampullary glands and seminal vesicles are connected with the ductus deferens, whereas the prostate and bulbourethral glands are connected with the urogenital sinus.

The differentiation, maintenance and functional integrity of these glands depends on androgen. Thus castration causes atrophy of secondary sexual organs. This atrophy is reflected by both involution of the epithelium and loss of secretion from the glands and tubules. The administration of testosterone will restore the functional activity and size of these structures. Condensations of the secretion are noted in the lumen of prostate of old men [20].

The urethral (Littré's) glands are multiple mucosal glands which open into the cavernous portion of the urethra. Some of these glands are simple outpocketings of the urethral mucosa formed by clear cells. Other urethral glands are somewhat larger and formed by a globular secretory portion connected with the urethra by a short duct [20]. Several preputial (Tyson's) glands are distributed all over the prepuce and on the neck of the penis. The histology of these glands resembles that of the sebaceous glands of the rest of the skin.

Conclusion

The transport of immotile spermatozoa from the testis is facilitated by the contractile activity of myoid cells, the flow of the fluid, and the movement of kinocilia.

Sertoli cells perform several functions: (a) they play a role in the coordination of spermatogenesis and metabolic exchange of germ cells before their release in the lumen of seminiferous tubules; (b) they have considerable phagocytic activity as shown by the presence of lysosomes, and (c) they have an endocrine function.

Leydig cells originate from fibroblasts during fetal life as well as prepubertal and postpubertal life. Elevation of testicular temperature as a result of cryptorchidism causes dramatic changes in testicular metabolic activity, e.g. decrease in glucose uptake, increase in metabolism of glucose to CO_2, and incorporation of lysine into protein [17].

Eight segments can be recognized in the epididymis as judged by the diameter of the lumen, the height and cytological characteristic of the epithelial cells, and cell organelles and cell surface.

During their transport in the epididymis, spermatozoa are exposed to differing biochemical milieu with varying concentrations of phosphorylcholine, carnitine, glutamic acid, sialic acids, steroids and a variety of enzymes.

The time of transit of spermatozoa through the epididymis varies from 12 to 21 days depending on the techniques employed for study. The differentiation, maintenance and functional integrity of these glands depends on androgen.

References

1 BAUMGARTEN, H.G.; OWMAN, C., and SJÖBERG, N.-O.: Neural mechanisms in male fertility; in SCIARRA, MARKLAND and SPEIDEL Control of male fertility, pp. 26–40 (Harper & Row, New York 1975).

2 BROBERG, A.; NYBELL, G.; OWMAN, C.; ROSENGREN, E., and SJÖBERG, N.-O.: Consequence of neonatal androgenization and castration for future levels of norepinephrine transmitter in uterus and vas deferens of the rat. Neuroendocrinology 15: 308–312 (1974).

3 COUROT, M.; HOCHEREAU-DEREVIERS, M.T., and ORTAVANT, R.: Spermatogenesis; in JOHNSON, GOMES and VANDEMARK The testis, pp. 339–432 (Academic Press, New York 1970).

4 DAVIS, J. and LANGFORD, G.A.: Pharmacological studies on the testicular capsule in relation to sperm transport; in ROSEMBERG and PAULSEN The human testis, pp. 495–514 (Plenum Press, New York 1970).

5 DE KRETSER, D.M.: The regulation of male fertility. The state of the art and future possibilities. Contraception 9: 561 (1974).

6 DE LA BALZE, F.A.; BUR, G.E.; SCARPA-SMITH, F., and IRAZU, J.: Elastic fibers in the tunica propria of normal and pathologic human testes. J. clin. Endocr. Metab. 14: 626 (1954).

7 DYM, M. and FAWCETT, D.W.: The blood-testis barrier in the rat and the physiological compartmentation of the seminiferous tubules. Biol. Reprod. 3: 308 (1970).

8 Hafez, E.S.E.: Human semen and fertility regulation in men (Mosby, St. Louis 1976).
9 Hafez, E.S.E.: Techniques of human andrology (Elsevier/North-Holland, Amsterdam 1977).
10 Hall, P.F.: Endocrinology of the testis; in Johnson, Gomes and Vandemark The testis, pp. 1–71 (Academic Press, New York 1970).
11 Holstein, A.F.: Morphologische Studien am Nebenhoden des Menschen; in Bargmann und Doerr Zwanglose Abhandlungen aus dem Gebiet der normalen und pathologischen Anatomie, vol. 20 (Thieme, Stuttgart 1969).
12 Hooker, C.W.: The intertubular tissue of the testis; in Johnson, Gomes and Vandemark The testis, pp. 483–550 (Academic Press, New York 1970).
13 Ionescu, B.; Duca-Marinescu, D., and Maximilian, C.: Genetic lesions of the Leydig tissue; in James, Serio and Martini The endocrine function of the human testis, pp. 67–99 (Academic Press, New York 1974).
14 Knight, T.W.: A qualitative study of factors affecting the contractions of the epididymis and ductus deferens of the ram. J. Reprod. Fert. $40:$ 19–29 (1974).
15 Lacy, D.: The seminiferous tubule in mammals. Endeavour $26:$ 101 (1967).
16 Lacy, D. and Pettit, A.J.: Sites of hormone production in the mammalian testis and their significance in the control of male fertility. Br. med. Bull. $26:$ 26 (1970).
17 Levier, R.R.: The influence of temperature on the incorporation of palmitic acid U-C^{14} into rat testicular lipids. J. Reprod. Fert. $15:$ 365 (1968).
18 Lostroh, A.J.: Hormonal control of spermatogenesis; in Spilman, Lobland and Kirton Regulatory mechanisms of male reproductive physiology (Excerpta Medica, Amsterdam 1976).
19 Monesi, V.: Spermatogenesis and the spermatozoa; in Austin and Short Reproduction in mammals. I. Germ cells and fertilization, pp. 46–84 (Cambridge University Press, Cambridge 1972).
20 Narbaitz, R.: Embryology, anatomy and histology of the male sex accessory glands; in Brandes Male accessory sex organs: structure and function in mammals, chapter 1 (Academic Press, New York 1974).
21 Rowley, M.J.; Teshima, F., and Heller, C.G.: Duration of transit of spermatozoa through the human male ductular system. Fert. Steril. $21:$ 390–396 (1970).
22 Vilar, O.: Histology of the human testis from neonatal period to adolescence; in Rosemberg and Paulsen The human testis, pp. 95–111 (Plenum Press, New York 1970).

Dr. E.S.E. Hafez, Reproductive Physiology Laboratories and Andrology Research Unit, C.S. Mott Center for Human Growth and Development, Wayne State University School of Medicine, *Detroit, Mich.* (USA)

Standards of Analysis and Clinical Biochemistry of Semen

R.F. Parrish, K.L. Polakoski and L.J.D. Zaneveld

Department of Obstetrics and Gynecology, Washington University School of Medicine, St. Louis, Mo. and Department of Physiology and Biophysics, University of Illinois at the Medical Center, Chicago, Ill.

Introduction

Several different trends in recent years have resulted in an increase in the number of semen analyses performed. Many infertility clinics include an evaluation of the ejaculate of the male partner as a routine clinical tool. The semen analysis is relatively simple and may eliminate lengthy and expensive gynecological testing of the female partner. The increased utilization of sperm banks and artificial insemination to achieve pregnancy require a semen analysis both before and after storage in order to insure that freezing has not damaged the spermatozoa. Finally, as vasectomy becomes a more accepted method of birth control, the number of analyses will increase proportionally, since the absence of spermatozoa in the ejaculate is the ultimate verification that the surgical procedure has been successful.

Semen is a viscous mixture of spermatozoa and seminal plasma. The spermatozoa are the major determinants of the man's fertilizing ability, although seminal plasma also plays a role. Seminal plasma is a complex mixture of fluids that serves as a vehicle for the transport of spermatozoa from the male to the female reproductive tract. Secretions from the testes, the epididymis and the accessory sex glands are emitted in a discrete sequence during ejaculation. The initial fraction, arising from the glands of Cowper and Littré, is thought to aid in the lubrication of the urethra. The second fraction is composed of prostatic fluid and the sperm-rich fluids derived from the distal vas and ampulla, and the epididymis. The final fraction is produced by the seminal vesicles. In a 'typical' ejaculate, the volumes of these three fractions are 0.2, 0.5, and 2.0–2.5 ml, respectively.

The fluids from the prostate and the seminal vesicles contribute different components to the total ejaculate. The concentration of these compounds are indicative of the proper functional states of their glands of origin. Thus, the information obtained from a semen analysis not only gives the physical characteristics of the spermatozoa, but also provides important information concerning dysfunctions of the accessory sex glands.

In this chapter, a description of the techniques for the collection of the ejaculate and the physical and biochemical tests to evaluate the semen sample are presented. Common problems associated with the analysis and possible physiopathological causes for observed abnormalities are included. An in-depth description of these procedures can be found in articles by POLAKOSKI and ZANEVELD [24] and ZANEVELD and POLAKOSKI [35].

Length of Abstinence before Semen Collection

The number of days to abstain from ejaculation before collection of a semen sample for analysis remains unresolved. Some believe that the abstinence period should be that normally observed by the patient. This has the advantage that the sample collected for analysis is representative of that deposited in the vagina during intercourse. Others, including ourselves, believe that the semen analysis should be performed on a specimen that represents the highest quality ejaculate which the patient is capable of producing.

It is well established that the length of abstinence before ejaculation causes significant changes in semen quality [18]. Although there are significant individual variations associated with the length of abstinence, there generally appears to be a significant and constant increase in sperm count during the first 4 days of continence. The increase in sperm numbers levels off thereafter, but gradual increases may still occur [29]. Abstinence for long periods is associated with an increase in sperm death and abnormal morphology [12], whereas short periods cause insufficient sperm maturation in the epididymis.

Optimally, a 3- to 4-day abstinence period should be observed for the initial sample collected for analysis. This period should not be less than 2 days nor more than 7 days under any circumstances. It is to be recommended that if the patient's normal abstinence period is substantially less than 4 days, or longer than 7 days, a second semen sample should be obtained at the patient's normal abstinence period. The results from this second sample should be compared to those obtained after a 4-day abstinence period. This will result in a more complete picture of the quality of the semen. At least one other sample, obtained 4–6 weeks later, should also be analyzed, since we have observed several instances where a sample contained very few spermatozoa, while several weeks later the same patient produced a normal ejaculate. The reasons for such wide divergence in sperm concentration from the same patient are not well understood, but probably result from a complex interrelationship between environmental, physiological, and psychological influences on the patient.

Collection of the Sample

The laboratory performing the analysis should provide the sample collection vessel. This avoids the problem of receiving samples in used and possibly contaminated food or medication containers. The opening of the sample container should be sufficiently large to accommodate entry of part of the glans penis and deep enough that the penis does not contact the bottom. The standard 2 ounce, wide-neck amber bottle (Boston Round) with a screw cap is excellent. The screw cap is strongly recommended since it does not easily come off in transit. The specimen containers must be scrupulously clean and dry since trace contamination by detergents are spermicidal, and water dilutes the sample and immobilizes spermatozoa. Plastic containers should be avoided, because some contain water-soluble spermicidal substances.

Masturbation is the method of choice for collection of the semen sample. Ideally, the sample should be obtained at the location where the analysis is to be performed. This minimizes delays in transport and variations in storage conditions. Unfortunately, this may not always be feasible. In such cases, the patient should bring the sample to the laboratory as soon after ejaculation as possible. Methods other than masturbation should be discouraged, although this may not be acceptable for religious or psychological reasons. The following alternate methods are listed in decreasing order of utility for the semen analysis.

A condom can be used to collect the ejaculate at the time of coitus. However, spillage occurs quite easily while removing the condom from the penis or while transferring the semen to the specimen container. Also, the rubber of most condoms is spermicidal. Whenever the condom method is employed, the patient should be advised to: (a) use nonspermicidal condoms (Milex Co.); (b) carefully remove the condom from the penis; (c) report any loss resulting from spillage or rupture of the condom.

Table I. Representative instructions for collecting the ejaculate

---

Instructions for collecting a semen sample
1. Abstain from ejaculation for 4 days.
2. Manual collection (masturbation) is the preferred method of semen collection. If another method is utilized, please note on the attached form.
3. The entire ejaculate is collected into the specimen bottle furnished by the laboratory. It is of critical importance to note on the attached form any loss of the ejaculate.
4. Bring the semen sample to the laboratory as rapidly as possible.
5. If the sample is collected at a site other than the laboratory performing the analysis, and the outside temperature is less than 70°F, carry the sample in an inner pocket and not in a briefcase or paper sack.
6. Fill in the information on the label and the attached form.

Semen can be collected by 'coitus interruptus'. Intercourse is performed as usual, but immediately before ejaculation, the penis is withdrawn and the semen deposited into the sample container. However, the first portion of the ejaculate is often lost. Since this portion contains the highest concentration of spermatozoa, its loss will lead to inaccuracies in the analysis.

Finally, semen samples can be obtained directly from the vagina after intercourse. A syringe, or tube with a suction bulb may be used to rinse the vagina with physiological saline immediately after intercourse. Alternatively, the vaginal or cervical fluids can be collected directly, without the aid of rinsing. All these samples are of questionable use, due to the small amounts of spermatozoa that can be obtained and the presence of female genital tract secretions that preclude meaningful biochemical analysis. Also, a significant time lapse usually occurs between the time of ejaculation and the collection of the fluids.

Both verbal and written instructions should be provided to the patient before the collection of the semen sample. All methods of collecting the semen sample, including the advantages and disadvantages of each method, should be discussed. Table I provides a representative set of instructions for collection of the sample by masturbation. Table II provides a representative form to be filled out by the patient at the time of presentation of the semen sample for analysis.

Table II. Representative information sheet to be presented with the semen sample

Sample Identification No. _____   Physician _____
Patient's Name _____
Address _____
Telephone No. _____
Date _____

1. Semen sample was collected by (check one)
   a) Masturbation _____
   b) Condom and intercourse _____
   c) Other (please specify) _____
2. Date and time of semen sample collection _____
3. Was any of the sample lost during collection?   YES _____ NO _____
4. If you answered YES to question 3, did the loss result from:
   a) Missing the container _____
   b) Spillage _____
5. If you checked 4a, which portion of the ejaculate was lost?
   a) First _____
   b) Middle _____
   c) Last _____
6. Semen sample was collected:
   a) At this location _____
   b) At another location _____
7. If you checked 6b, approximately how far was the sample transported? _____
8. Was any of the sample spilled while bringing the sample to the laboratory?   YES _____ NO _____
9. If you answered YES to question 8, how much of the sample was lost?   25% ___ 50% ___ 75% ___ 100% ___
10. Number of days since your last ejaculation _____
11. Number of times you ejaculate (intercourse, masturbation)   _____ times/week
    _____ times/month

To be Completed by the Laboratory when the Sample Arrives

12. Date and time of arrival _____
13. Any apparent loss of sample?   YES _____ NO _____
14. Comments: _____

Transportation of the Sample

If a sample cannot be obtained at the laboratory where the analysis is performed, every effort should be made to impress the patient with the importance of rapidly transporting the sample to the laboratory. If the sample is transported during warm weather, it should be maintained at ambient temperature. In cold weather, the patient should be advised to keep the sample in an inner pocket and never in a briefcase or exterior pocket exposed to the cold. If delays of more than 4 h are unavoidable, the sample should be stored in a refrigerator. It is imperative that the specimen is not and has not been frozen and thawed, because this will immobilize a high percentage of the spermatozoa. Likewise, the sample should never be heated above body temperature, since sperm damage will occur, as well as changes in semen biochemistry.

The results obtained from a sample received more than 60 min after ejaculation are suspect. If more than 2 h have elapsed since ejaculation, definite changes in the sample will have occurred and if the sample is abnormal in any way, another sample must be obtained and the time interval between ejaculation and arrival at the laboratory must be minimized.

Sample Preparation

After liquefaction has occurred and the volume has been measured, the sample is divided into roughly equal parts. One portion is used for the physical examination, and the other for the biochemical tests. The latter portion is centrifuged at $700\,g$ for 15 min to sediment the spermatozoa. The removal of the spermatozoa is necessary since spermatozoa metabolize fructose, one of the compounds measured in the biochemical tests. The sperm-free supernatant is maintained at $0\,°C$ or frozen if the biochemical tests cannot be performed immediately. Table III shows a representative form for reporting the results of the semen analysis.

Physical Examination of the Ejaculate

Coagulation and Liquefaction

Immediately after arrival, the specimen is observed for the presence of a thick viscous mass that is difficult to draw into a pipette. This is the result

Table III. Representative form for reporting the results of the semen analysis

Patient's Name _____   Sample Identification No. _____

Date _____   Physician _____

Date of last ejaculation _____

Test Performed	Normal	
	YES	NO
1. Coagulum present YES _____ NO _____	1. _____	_____
2. Time for complete liquefaction _____ minutes	2. _____	_____
3. Normal odor	3. _____	_____
4. Color _____	4. _____	_____
5. Volume _____ ml	5. _____	_____
6. pH _____	6. _____	_____
7. Viscosity – Normal _____ High _____	7. _____	_____
8. Sperm concentration _____ sperm/ml	8. _____	_____
9. Sperm count _____ sperm	9. _____	_____
10. Per cent viable sperm _____ %	10. _____	_____
11. Per cent motile spermatozoa	11. _____	_____

Hours after ejaculation   1. ____%
                          2. ____%
                          3. ____%
                          4. ____%
                         24. ____%

12. Per cent motile spermatozoa with forward progression   12. _____ _____

Hours after ejaculation   1. ____%
                          2. ____%
                          3. ____%
                          4. ____%
                         24. ____%

13. If the per cent of viable sperm is substantially higher than the percentage of non-motile sperm, the following tests are performed and observed for increased motility:

	Increased Motility	
a) Sugar addition	YES____	NO____
b) Caffeine addition	YES____	NO____
c) Arginine addition	YES____	NO____
d) Kallikrein addition	YES____	NO____

Table III. (continued)

14. Sperm agglutination	___ None ___ Slight ___ Moderate ___ Heavy	14. ___	___
15. Per cent spermatozoa with abnormal morphology (Indicate the type of abnormality if one or more types predominate)	___ %	15. ___	___
16. Fructose	___ mg/ml	16. ___	___
17. Citric Acid	___ mg/ml	17. ___	___
18. Acid Phosphatase	___ IU/ml	18. ___	___
19. Comments			
		Signature ___ Date ___	

of the seminal coagulation, which normally liquefies in 15–20 min. The time required for complete liquefaction should be recorded.

The seminal vesicles supply the coagulating proteins, but the liquefaction agents, probably enzymes, are products of the prostate [31]. Congenital aplasia of seminal vesicles or an obstruction of the ejaculatory duct is associated with the absence of the coagulum. A prostatic dysfunction is often indicated by a liquefaction time longer than 1 h. Patients with poorly lysing ejaculates may be subfertile, since a high percentage of the spermatozoa are trapped by the coagulum and cannot progress through the female genital tract [16].

For most of the other physical tests, a liquified sample is required and several techniques are available to artificially induce liquefaction. Proteolytic enzymes such as trypsin [26] or a 5% solution of α-amylase [4] have been employed. Alevaire in a 1:1 ratio with semen works quite well [1]. The viscosity can also be reduced by filtering the samples through a glass wool column [22]. In this case, the dead and agglutinated spermatozoa as well as the debris are removed from the sample. It must be kept in mind that such procedures may change the properties of the sample. If an artificial method of liquefaction is utilized, it must be noted on the patient's records. Artificially liquified samples may be used for AIH (artificial insemination of husband semen) if the lack of liquefaction appears to be the cause of infertility.

Odor

Normal semen has a strong and pungent odor thought to be the result of spermine oxidation [21]. Spermine is produced by the prostate gland and lack of a strong odor may indicate a prostatic dysfunction [11].

Color

When observed against a white background the normal ejaculate is translucent or whitish-gray. A white or yellow appearance can be caused by the presence of large numbers of white blood cells, but a yellow color may also result from prolonged sexual abstinence. Red discolorations indicate bleeding of the tract (hemospermia). Antibiotics used to treat cystitis may also cause discoloration of the ejaculate.

Volume

A pipette, a small graduated cylinder or a graduated centrifuge tube is used to measure the volume of the ejaculate to the nearest 0.1 ml. A patient who produces no semen on ejaculation is aspermic. One who produces less than 1.0 ml is hypospermic, while one who produces more than 6.0 ml of semen is hyperspermic. In cases of low volume, it is essential to ascertain if the sample presented for analysis is the entire ejaculate, or if a portion of the ejaculate was lost due to spillage or poor sample collection. If part of the ejaculate was lost by missing the sample container, this will drastically affect the subsequent physical and biochemical tests. A marked reduction in the sperm count and concentration will result if the first portion of the ejaculate is lost, while conversely, a high sperm concentration, but normal sperm count will result if the final portion of the ejaculate is lost. Loss of the first portion of the ejaculate may also result in poor liquefaction.

The congenital absence of seminal vesicles, or an obstruction of the ejaculatory duct results in decreased volumes and the absence of spermatozoa. Low volumes in the presence of spermatozoa or sperm precursors may indicate a pituitary gland or a testicular interstitial (Leydig) cell deficiency. Volumes greater than 6.0 ml can result from overactivity of the accessory sex glands, or may simply be due to a long period of sexual abstinence. High volumes are normally associated with subnormal sperm concentrations, although the total sperm count may be normal. Such high volumes may interfere with sperm transport in the female or may contribute to loss of spermatozoa from the vagina by a 'washing out' process and consequently result in reduced fertility.

pH

The pH of the semen sample is measured immediately after the liquefaction, since pH changes occur thereafter [25]. An electric pH meter is preferred, but placing a drop of semen on pH paper with a range of 6.6–8.0 is satisfactory. The normal pH range is 7.2–7.8 and results from a mixture of the acidic prostatic fluid and the basic seminal vesicle fluid. Acute inflammatory diseases of the accessory sex glands or of the epididymis will result in a pH above 8.0. Dysfunction or congenital aplasia of the seminal vesicles, or an obstruction of the ejaculatory duct result in a pH lower than normal. If the first portion of the ejaculate is lost, the resulting sample will have a pH higher than normal, while loss of the final portion of the ejaculate will cause the sample to have a pH lower than normal.

Viscosity

A small amount of semen is drawn into a Pasteur pipette and the degree of difficulty for this is noted. As the pipette is removed from the sample, a small thread of semen adheres to the pipette tip. This thread can be 3–5 cm long, even after complete liquefaction of the coagulum. The greater the viscosity of the semen sample, the longer the thread. Finally, the specimen container is rotated and the ability of the semen sample to conform to the sample container is noted. A description of the viscosity is noted, i.e., normal more viscous than normal, etc. Unfortunately, these measurements are very subjective and rely heavily on the experience of the observer.

A more complicated, but also more accurate method, utilizes the time required for a drop of semen to leave a calibrated pipette [7]. A pipette possessing a volume of 0.1 ml with a length of 12 cm between the tip and the 0.1 cm mark is standardized against silicon oil MS 200, viscosity 20 cSt. Three drops of oil leave the pipette in 5 ± 0.2 sec. To measure the viscosity of the semen sample, the semen is drawn into the pipette and after the pressure is released, the time necessary for one drop of semen to form is noted. If it requires more than 5 sec for a drop of semen to leave the pipette, the sample is abnormally viscous.

High viscosity is not synonymous with poor liquefaction. The two phenomena can be distinguished by rotating the sample container. A highly viscous sample will slowly conform to the shape of the sample container, but a sample that has not liquefied will not conform to this shape. Spermatozoa trapped in a highly viscous ejaculate are immobilized and therefore unable to begin passage through the female tract, resulting in diminished fertility.

Sperm Count and Concentration

The sperm count is the total number of spermatozoa in the ejaculate, while the sperm concentration is the number of spermatozoa per milliliter of semen. The standard method of counting spermatozoa consists of killing the spermatozoa and then counting the dead spermatozoa utilizing the blood cell hemocytometer and a microscope at 100–400×. The sperm are killed since it is otherwise difficult to count spermatozoa that are moving. The immobilizing agents also help to break up aggregated spermatozoa. The commonly employed spermicides are: (a) 5% triphenyltetrazolium chloride in physiological saline; (b) 5% chlorazene in physiological saline; (c) 5 g of sodium carbonate and 1 ml of 35% formaldehyde diluted to 100 ml with physiological saline. A 20-fold dilution of semen with spermicide is standard. This is accomplished by adding 0.05 ml of semen to 0.95 ml of spermicide. However, for semen samples with substantially higher or lower than normal sperm concentrations, much larger or smaller dilutions should be made. After the appropriate dilution of the sample with spermicide and thorough mixing, one drop is added to both sides of the hemocytometer.

A grid system divides the hemocytometer in five major zones, designated A, B, C, D and E. The central E square is further divided into 25 small squares with the corner squares designated E_1, E_2, E_3, E_4 the central square designated E_5. Either all the spermatozoa in the major square E are counted, or only the spermatozoa in the small squares E_1, E_2, E_3, E_4, and E_5 are counted. The count should include all spermatozoa within the designated square or squares, and also those spermatozoa that cross the double lines on two sides, i.e., at the top and right hand side of each square. Both sides of the hemocytometer are counted, and the results averaged. The accuracy of this method is not extremely high. Differences of 20–30% are to be expected when duplicate determinations are performed by the same person. It has been claimed that dilutions of 1:100 or 1:200 result in a higher degree of accuracy than can be obtained by the standard 1:20 dilution. It must be kept in mind, however, that when dealing with these larger dilutions, the sample must be thoroughly mixed, since small differences in the sampling are magnified by the dilution. Large variations in the sperm count occur when the sample is highly viscous, for the aliquot removed from the sample is not necessarily from an evenly distributed sample. The authors have seen 10-fold variations in the sperm count from such highly viscous samples.

In order to determine the sperm concentration and total sperm count, several terms must be defined. The volume of the sample represented by the major square E of the hemocytometer is 1×10^{-4} ml. Therefore, to obtain the number of spermatozoa applied to the hemocytometer, the number of sperm in the major square E must be multiplied by 10,000. 10,000 is then defined as the multiplication factor. If only squares E_1, E_2, E_3, E_4, and E_5 are counted, the volume represented by these five squares is only one fifth of square E, that is, 2×10^{-5} ml, and the multiplication factor is 50,000. In a similar manner, if the spermatozoa in squares A, B, C, D, and E are counted, this represents 5 times the volume of square E and the multiplication factor is 2,000. The dilution factor is defined as the amount that the initial semen sample was diluted before application to the hemocytometer. For a 1:20 dilution, the dilution factor is 20. Sperm concentration in sperm per milliliter and total sperm count are calculated from the following equations:

Sperm concentration = (number of spermatozoa counted) × (the multiplication factor) × (the dilution factor).

Total sperm count = (sperm concentration) × (volume of the semen sample).

Normospermia, polyzoospermia, and oligozoospermia are respectively defined as concentrations of 20–250, greater than 250, and less than 20 million spermatozoa per milliliter. Azoospermia is defined as the complete absence of spermatozoa in the ejaculate. The standard method is not satisfactory for diagnosing azoospermia and the following modification is utilized. The semen sample is centrifuged at 2,000 g for 15 min and the supernatant seminal plasma is removed. The precipitated material is removed and added to a slide and carefully observed microscopically for the presence of spermatozoa. Several ejaculates must be so observed before azoospermia can be unequivocally confirmed.

Motility and Forward Progression

Motility is temperature dependent [13]. This can cause difficulties, since it is frequently difficult to establish what temperatures spermatozoa have been exposed to during transport. Also, there are wide variations in 'room temperature' between different laboratories, and even within different areas of the same laboratory, In order to minimize the chances for misinterpretation, all semen samples should be warmed at 37°C, the temperature of the spermatozoa in their natural environment, before evaluation of motility.

After liquefaction of the coagulum, either naturally or through the use of an exogenous lytic agent, the sample is thoroughly mixed and a drop of undiluted semen is placed on a slide and covered with a cover slip. The 'hanging drop' method, where the drop of semen is placed on a cover slip and then placed upside down on a slide containing a central depression, can also be used. The semen can be transferred with a bacteriological loop that delivers constant amounts of fluid and has the advantage that it can be sterilized after each application or a pipette can be used. After 10 min on the slide stage at 37 °C, the spermatozoa are viewed with a microscope at 100 or 400 ×. A minimum of 200 spermatozoa are counted in at least four randomly selected fields. The percent motility is calculated by dividing the number of spermatozoa with any form of movement by the total number of spermatozoa present. The percent of spermatozoa with forward progression is calculated by dividing the number of spermatozoa with forward progression by the number of motile spermatozoa. Spermatozoa that move in small circles while essentially remaining in the same area are considered to have stationary movement.

A sample can be considered normal when, after 1 h postejaculation, 70% of the spermatozoa show movement, 70% of this being progressive. A sample with either less than 50% motility or less than 50% progressive movement of the motile sperm is considered abnormal. A significant drop in motility after 2–4 h also indicates an abnormal sample. Observations should therefore be made every hour for 4 h and again after 24 h. It should also be noted if white blood cells or large numbers of bacteria are present, for this not only indicates a genitourinary infection, but many types of bacteria secrete sperm immobilizing factors that adversely affect sperm motility [23]. Other factors that affect spermatozoa motility include: heat, cold shock, many plastics, detergents, improper cleaning of specimen bottles, microscope slides and cover slips, long periods of sexual abstinence or even physiological and psychological factors. The agglutination of spermatozoa may also influence the estimation of sperm motility and progression. The motility is frequently normal, but the forward progression will be poor. We consider spermatozoa to be agglutinated only when several aggregates of four or more spermatozoa are adhering to each other, usually in a head to head or tail to tail manner. It is imperative to report sperm agglutination, since this usually indicates the presence of sperm antibodies. However, this cannot be unequivocally deduced unless further tests are performed, since spermatozoa also have a tendency to adhere to cellular debris, bubbles in the sample, the air-sample interface, and white blood cells.

Live-Dead (Supravital) Stain

It is possible that nonmotile spermatozoa are not dead and under certain circumstances may regain motility. Therefore, it is important to distinguish between dead spermatozoa and nonmotile spermatozoa. Dead spermatozoa are characterized by damaged head membranes and this damage can be identified by the penetration of the sperm head by eosin, utilizing one of the following methods:

(A) 0.1 ml of semen is mixed with 0.1 ml of eosin Y (0.5% in 0.15 M phosphate, pH 7.4), and after 1–2 min a smear is prepared and air dried. The smear is observed microscopically using negative phase contrast and oil immersion (1,000–1,250 ×). Dead spermatozoa are bright yellow, while viable spermatozoa are stained bluish [8].

(B) One drop of semen is mixed with 1 drop of bluish eosin (0.5% in distilled water), 2 drops of Nigrosin (10% in distilled water) are added, and the sample thoroughly mixed. A sperm smear is prepared, air dried, and viewed under the microscope at 400 × or under oil immersion at 1,000–1,250 ×. Dead spermatozoa are stained red against a red background, while viable spermatozoa will be unstained [3].

In order to avoid confusion from partially stained spermatozoa, if any part of the spermatozoon shows the color of death, it should be considered dead. At least 200 spermatozoa should be counted and the percent viable is calculated. The percent dead spermatozoa should always be equal to or less than percent of spermatozoa found to be nonmotile. If the percentage of nonmotile spermatozoa is significantly higher than the percent of dead spermatozoa, either immobilizing agents are present, or the nutrient level of the semen is subnormal. In either case, the tests presented in the following section should be performed.

Stimulation of Sperm Motility and/or Forward Progression

Spermatozoa that are nonmotile may possess the potential for motility once they are in the female genital tract. The following *in vitro* tests serve to discriminate nonmotile and potentially motile spermatozoa. At least two should be performed.

Sugar stimulation. Equal volumes of semen are mixed with Baker's solution (3 g glucose, 0.6 g Na_2HPO_4, 0.01 g K_2HPO_4, 0.2 g NaCl, add distilled water to a final volume of 100 ml), Joel's solution (80 ml of 5.42% dextrose in distilled water, 20 ml of 0.125 N $MgCl_2$ in distilled water), or Locke's solution (0.024 g $CaCl_2$, 0.042 g KCl, 0.01 g $Na.CO_3$, 0.9 g NaCl, 0.1 g glucose, add distilled water to a final volume of 100 ml), incubated at 37°C and the motility and forward progression compared every hour for 4 h to a sample prepared by mixing equal volumes of semen and physiological saline (0.85 g NaCl, add distilled water to a final volume of 100 ml) [29].

Caffeine stimulation. 0.1 ml of modified Ringer's buffer solution (0.7 g NaCl, 0.037 g KCl, 0.15 g KH_2PO_4, 0.06 g $MgCl_2$, 0.02 g Tris · HCl, add distilled water to a final volume of 100 ml) containing 36 mM caffeine is added to 0.5 ml of semen, incubated at 37°C and the motility and forward progression are compared every hour for 4 h to a sample prepared by mixing 0.1 ml of modified Ringer's solution with 0.5 ml of semen [30].

Arginine stimulation. The semen is diluted with physiological saline to a concentration of 20×10^6 spermatozoa/ml. An equal volume of semen and L-arginine (8 mM in physiological saline) is mixed and incubated at 37°C. The sperm motility and forward progression are compared every hour for 4 h to semen incubated with physiological saline [14].

Kallikrein stimulation. 0.25 ml semen is mixed with 0.25 ml of physiological saline containing 3×10^{-8} M kallikrein, incubated at 37°C and the sperm motility and forward progression are compared every hour for 4 h to a sample prepared with equal volumes of semen and physiological saline [28].

Morphology

Examination of the morphology of spermatozoa can only be performed by staining. For samples of normal sperm count, a semen smear is prepared on a slide and air dried. When the sperm count is less than 20 million/ml the sample is centrifuged for 15 min at 2,000 g and a semen smear is prepared from the sedimented material and air dried. Fixing and staining is accomplished by layering 10% formaldehyde on the slide incubating for 1 min at room temperature and rinsing with distilled water. The slide is covered with Meyer's hematoxylin, incubated for 2 min at room temperature and rinsed with distilled water [10]. Other methods are available that result in greater contrast for the individual sperm structures, but they are much more complicated. The smear is observed under the microscope at 400 × or under oil immersion at 1,000–1,250 ×. A minimum of 200 spermatozoa are counted. The percent normal and abnormal forms, the type of abnormality, and the presence of other cells, i.e., white blood cells, bacteria, are recorded.

Wide variations in 'normal' spermatozoa are found. This means that determination of abnormal forms is, to a large extent, dependent on the experience of the observer. The inexperienced observer should consult a chart for comparisons between the normal and abnormal forms [9]. Two common problems encountered in defining abnormality result from differences in the angle between the plane of the spermatozoa, and the observer. Since the head of the spermatozoa has a flattened oval appearance, when the spermatozoa is laying flat with respect to the viewer, the head will appear rounded. However, when the spermatozoon is viewed from the side, the head will seem pointed. Intermediate positions are possible, with the resulting differences in the appearance of the head. Superimposition of two spermatozoa may give the impression of two heads or tails. Tail abnormalities may be difficult to determine since the tail is normally bent to provide sperm movement. An exaggerated coil or distinctly angular appearance should be present before reporting a tail abnormality.

Up to 30–40% of all spermatozoa present in human semen possess an abnormal morphology [6]. When this percentage increases to greater than

50%, infertility often occurs. Such observations must be considered as indicators and not complete diagnostic tools, since the reasons for abnormal morphology are manifold and can also be caused by physical or psychological stress. The presence of large numbers of abnormal spermatozoa indicates a testicular abnormality. Predominance of one or more abnormalities can sometimes be indicative of the cause of the abnormality. The presence of tapered heads and sperm precursors are frequently the result of viral or bacteriological infection. This is almost certain in the presence of large numbers of white blood cells. The presence of a large percentage of tapered headed spermatozoa is often indicative of varicocele [17]. An increase in the amorphous spermatozoa and sperm precursors often indicates an allergic reaction [17]. An epididymal dysfunction is indicated by the presence of large numbers of spermatozoa containing cytoplasmic droplets, although this may also be the result of frequent ejaculations. Collection of a semen sample after a longer period of abstinence should discriminate between these two states.

Biochemical Analysis of the Ejaculate

At the present time, there is no definitive biochemical test that can discriminate between a fertile and an infertile male. However, when combined with the physical examination of the ejaculate, the biochemical analysis yields important information concerning the status of the male accessory glands. Prostatic fluid is slightly acidic and is the source of zinc, citric acid phosphatase, and the seminal clot liquefying agents [20]. The absence or reduced levels of these agents is indicative of a lack of prostatic fluid. This may result from an obstruction of the prostatic duct as caused by inflammation, benign prostatic hypertrophy, or adenocarcinoma.

Seminal vesicle secretions provide the seminal clot forming proteins and fructose [19]. Failure of the ejaculate to coagulate, a pH of the ejaculate below 6.7 or the absence of fructose indicates the absence of seminal vesicle secretions. The presence or absence of fructose can be utilized to determine if azoospermia results from a testicular abnormality, or the absence or blockage of the vas deferens. The vas deferens and the seminal vesicles are derived embryologically from the Müllerian duct. In the absence or blockage of the seminal vesicles, there will be no spermatozoa or fructose in the ejaculate. However, if the absence of spermatozoa only results from a testicular abnormality, fructose will be present [12].

The analyses presented in this section all utilize a standard curve to estimate the concentration of an unknown substance of interest, and rely on a colorimeter or a spectrophotometer to obtain the raw data. Analyses should be performed in duplicate and it is strongly recommended that two samples of known concentration be included in each series of determinations. If differences are observed between experimentally determined concentrations and the known concentration of the standards, this indicates that either an error has been made in the experimental technique, or one or more of the reagents employed in the analysis has deteriorated. Whenever new reagents are prepared, a new standard curve must also be prepared.

Fructose

(1) 0.1 ml of seminal plasma is added to 2.9 ml of distilled water and 0.5 ml of 5% zinc sulfate and 0.5 ml of 0.3 N barium hydroxide are added and thoroughly mixed. The sample is heated at 100°C for 1 min and centrifuged at 700 g for 15 min.

(2) 2.0 ml of the deproteinized solution from step 1 is thoroughly mixed with 2.0 ml of 0.1% resorcinol (0.1 g dissolved in 100 ml of 95% ethanol) and 6.0 ml of 30% hydrochloric acid (50 ml of concentrated hydrochloric acid added to 10 ml of water). The solution is heated in a water bath for 10 min at 90°C and cooled to room temperature.

(3) A blank is prepared by replacing the seminal plasma with water and is treated in an identical manner.

(4) The optical density of the sample and the blank are measured at 410 nm within 30 min. The blank value is subtracted from the sample value and the amount of fructose present is determined from a standard curve (10–200 mg) prepared from commercially available fructose [21].

The normal range of values is 0.7–5.0 mg/ml of semen, with a mean of 3.0 mg/ml [32]. This procedure measures not only the free fructose, but also the phosphofructose. Spermatozoa metabolize fructose as an energy source. It is therefore imperative that spermatozoa be separated from the seminal plasma as soon after liquefaction as possible.

Citric Acid

(1) 1.0 ml of seminal plasma is added to 1.0 ml of 50% trichloroacetic acid (50 g trichloroacetic acid in 100 ml total volume of water) and after chilling for 15 min on ice, the precipitate is removed by centrifugation at 700 g for 15 min. If less than 1.0 ml of seminal plasma is used, an equal volume of seminal plasma and 50% trichloroacetic acid should be mixed and diluted to 2.0 ml with 25% trichloroacetic acid.

(2) 8.0 ml of anhydrous acetic anhydride (reagent grade) is added to 1.0 ml of the deproteinized sample. The tube is securely stoppered and heated at 60°C for exactly 10 min, at which time 1.0 ml of dry pyridine (reagent grade) is added. After recapping, the tube is heated at 60°C for another 40 min and then chilled in an ice bath for 5 min.

(3) A blank is prepared utilizing 1 ml of 25% trichloroacetic acid and is treated in an identical manner.

(4) The optical density of the sample and the blank are measured at 400 nm within 1 h. After subtraction of the blank value from the sample value, the concentration is obtained from a standard curve (25–400 µg) [27].

The chromophore produced by this method is very temperature sensitive. It is essential that the temperature be maintained at 60±1 °C. The range of reported normal values is 1.8–8.4 mg/ml with a mean of 5.1 mg/ml [5].

Acid Phosphatase

(1) Seminal plasma is diluted with physiological saline (1:100–1:1,000) and 0.2 ml is added to 1.0 ml of *p*-nitrophenol phosphatase (2.75 mM in 90 mM citrate, pH 4.8) that was preincubated at 37°C for 5 min. After exactly 30 min at 37°C, the sample is removed from the bath and 5.0 ml of 0.1 N sodium hydroxide is added.

(2) A blank is prepared utilizing physiological saline instead of seminal plasma and is treated in exactly the same manner.

(3) The optical density is measured at 410 nm and after subtraction of the blank value from the sample value, the amount of nitrophenol liberated is obtained from a standard curve. The standard curve was prepared by diluting a stock solution of *p*-nitrophenol (10 µmol/ml in distilled water) with 0.02 N sodium hydroxide to a range of 5–50 nmol/ml. The normal range is 88–979 IU/ml with a mean value of 408 IU/ml [6]. 1 IU is defined as the amount of enzyme that yields 1 µmol *p*-nitrophenol/min.

Spermatozoa Penetration of Cervical Mucus

A simple test has been developed to determine the ability of spermatozoa to penetrate cervical mucus [15]. This test is a good supplement to the normal semen analysis, since failure of spermatozoa to penetrate the cervical mucus will result in infertility. A sperm penetration meter is prepared by splitting a 70×7 mm test tube lengthwise and cutting off the bottom 1 cm for each split tube. This bottom piece is glued to a glass slide to provide a semen reservoir. A thin glass rod is glued to the slide approximately 1 cm from the mouth of the semen reservoir and the slide is calibrated at 1-cm intervals with etching ink. Cervical mucus is obtained from a donor at the time of ovulation and should: (a) be clear, (b) have low viscosity, (c) have a spinnbarkeit of at least 50 mm, (d) exhibit a positive fern test, and (e) demonstrate penetration of at least 40 mm in 1 h by spermatozoa from a fertile donor of known cervical mucus penetration ability [34]. The cervical mucus is drawn into a capillary tube to a height of 7 cm, sealed at both ends with modeling clay and stored in the refrigerator until use. Capillary tubes prepared in this manner are usable for several weeks. Liquefied semen is placed in the semen reservoir. The modeling clay is removed from one end of the capillary tube and the modeling clay at the other end is pushed into the capillary tube forcing a small drop of mucus to protrude from the exposed end. This end is placed in the semen sample and is supported by the glass rod and the modeling clay. The entire apparatus is then incubated in a moist Petri dish at 37 °C and examined under the microscope every half hour for 3 h. The distance the foremost spermatozoa have penetrated is recorded. Normal spermatozoa penetrate at least 20 mm into the cervical mucus after 3 h at 37 °C [33].

Conclusions

Impregnation of the female is the ultimate verification of fertility in the male. Unfortunately, there exists no definite test that will establish male infertility. The semen analysis provides the best available data to determine the apparent normality of the male ejaculate, and by analogy, the fertilization potential of the patient. This is accomplished by comparing the patient's semen to an empirically determined set of 'normal' values obtained from a multitude of analyses of the semen of males of known fertility. Gross deviations of the patient's semen from the normal values indicate a 'possible' reason for infertility, but cannot be taken as a definitive diagnosis, since the literature is replete with reported cases of fertility of males whose semen analysis indicated a subnormal quality in one or more semen parameters. It is important to remember that an abnormal semen parameter is a result of an abnormality in the male reproductive system and is not a causative agent in and of itself. It is the job of the andrologist to determine the reason for the abnormality and to prescribe a method of treatment. The monitoring of this method of treatment will be the semen analysis and the observation of an improved semen sample.

References

1 AMELAR, R.D.: Coagulation, liquefaction and viscosity of human semen. J. Urol. *87:* 187–190 (1962).
2 AMELAR, R.D. and HOTCHKISS, R.S.: Congenital aplasia of the epididymes and vasa deferentia: effects on semen. Fert. Steril. *14:* 44–48 (1963).
3 BLOM, E.: A one-minute live-dead sperm stain by means of eosin-nigrosin. Fert. Steril. *1:* 176–177 (1950).
4 BUNGE, R.G. and SHERMAN, J.K.: Liquefaction of human semen by alpha-amylase. Fert. Steril. *5:* 353–356 (1954).

5 DONDERO, F.; SCIARRA, F., and ISIDORI, A.: Evaluation of relationship between plasma testosterone and human seminal citric acid. Fert. Steril. *23:* 168–171 (1972).
6 ELIASSON, R.: Standards for investigation of human semen. Andrologie *3:* 49–64 (1971).
7 ELIASSON, R.: Parameters of male fertility; in HAFEZ and EVANS Human reproduction. Conception and contraception, pp. 39–51 (Harper & Row, New York 1973).
8 ELIASSON, R. and TREICHL, L.: Supravital staining of human spermatozoa. Fert. Steril. *22:* 134–137 (1971).
9 FREUND, M.: Standards for the rating of human sperm morphology. Int. J. Fert. *11:* 91–180 (1966).
10 FREUND, M. and PETERSON, R.N.: Semen evaluation and fertility; in HAFEZ Human semen and fertility regulation in men, pp. 344–354 (Mosby, St. Louis 1976).
11 HARRISON, G.A.: Spermine in human tissues. Biochem. J. *25:* 1885–1892 (1931).
12 HARVEY, C. and JACKSON, M.H.: Assessment of male fertility by semen analysis. Lancet *ii:* 99–102 (1945).
13 JANICK, J. and MACLEOD, J.: The measurement of human spermatozoa motility. Fert. Steril. *21:* 140–146 (1970).
14 KELLER, D.W. and POLAKOSKI, K.L.: L-Arginine stimulation of human sperm motility *in vitro*. Biol. Reprod. *13:* 154–157 (1975).
15 KREMER, J.: A simple sperm penetration test. Int. J. Fert. *10:* 209–215 (1965).
16 MACLEOD, J.: Biochemistry of the male genital tract. Ann. N.Y. Acad. Sci. *54:* 796–805 (1951).
17 MACLEOD, J.: Human seminal cytology following the administration of certain antispermatogenic compounds; in AUSTIN and PERRY Agents affecting fertility, pp. 93–123 (Little, Brown, Boston 1965).
18 MACLEOD, J. and GOLD, R.Z.: The male factor in fertility and infertility. V. Effect of continence on semen quality. Fert. Steril. *3:* 297–315 (1952).
19 MANN, T.: Studies on the metabolism of semen. 3. Fructose as a normal constituent of seminal plasma, site of formation and function of fructose in semen. Biochem. J. *40:* 481–491 (1946).
20 MANN, T.: Biochemistry of the prostate gland and its secretions; in Biology of the prostate and related tissues. Natn. Cancer Inst. Monogr., No. 12, pp. 235–251 (US Department of Health, Education and Welfare, Bethesda 1963).
21 MANN, T.: The biochemistry of semen and of the male reproductive tract, pp. 193–264 (Methuen, London 1964).
22 PAULSON, J.D. and POLAKOSKI, K.L.: A glass wool column procedure for removing extraneous material from the human ejaculate. Fert. Steril. *28:* 178–181 (1977).
23 PAULSON, J.D. and POLAKOSKI, K.L.: Isolation of a spermatozoal immobilization factor from *Escherichia coli* filtrates. Fert. Steril. *28:* 182–185 (1977).
24 POLAKOSKI, K.L. and ZANEVELD, L.J.D.: Biochemical examination of the ejaculate; in HAFEZ Techniques in human andrology, pp. 265–286 (Elsevier, New York 1977).
25 RABOCK, J. and SHACHOVÁ, J.: The pH of human ejaculate. Fert. Steril. *16:* 252–256 (1965).
26 ROUSSEL, J.D. and AUSTIN, C.R.: Enzymic liquefaction of primate semen. Int. J. Fert. *12:* 288–290 (1967).
27 SAFFRAN, M. and DENSTEDT, O.F.: A rapid method for the determination of citric acid. J. biol. Chem. *175:* 849–855 (1948).

28 SCHILL, W.B.: Caffeine and kallikrein induced stimulation of human sperm motility: a comparative study. Andrologia 7: 229–236 (1975).
29 SCHIRREN, C.: Practical andrology, pp. 19–24 (Hartmann, Berlin 1972).
30 SCHOENFELD, C.; AMELAR, R.D., and DUBIN, L.: Stimulation of ejaculated human spermatozoa by caffeine. Fert. Steril. 26: 158–161 (1975).
31 TAUBER, P.F.; PROPPING, D.; ZANEVELD, L.J.D., and SCHUMACHER, G.F.B.: Biochemical studies on the lysis of human split ejaculates. Biol. Reprod. 9: 62 (1973).
32 TAUBER, P.F.; ZANEVELD, L.J.D.; PROPPING, D., and SCHUMACHER, G.F.B.: Components of human split ejaculates. I. Spermatozoa, fructose, immunoglobulins, albumin, lactoferrin, transferrin and other plasma proteins. J. Reprod. Fert. 43: 249–267 (1975).
33 ULSTEIN, M.: Evaluation of a capillary tube sperm penetration method for fertility investigations. Acta obstet. gynec. scand. 51: 287–292 (1972).
34 ULSTEIN, M.: Sperm penetration of cervical mucus as a criterion of male fertility. Acta obstet. gynec. scand. 51: 335–340 (1972).
35 ZANEVELD, L.J.D. and POLAKOSKI, K.L.: Collection and physical examination of the ejaculate; in HAFEZ Techniques in human andrology, pp. 147–172 (Elsevier, New York 1977).

Dr. R.F. PARRISH, Department of Obstetrics and Gynecology, Washington University School of Medicine, *St. Louis, MO 63110* (USA)

Neuroendocrine Parameters of Male Fertility and Infertility

Jerald Bain

Departments of Medicine and Obstetrics and Gynecology, University of Toronto, and Reproductive Biology Unit, Mount Sinai Hospital, Toronto, Ont.

A considerable body of new knowledge regarding male reproductive processes has appeared within recent years. The development of sensitive and specific radioimmunoassays for both pituitary hormones and sex steroids has allowed for the measurement of extremely small concentrations of circulating hormones which in turn has provided information to explain many physiologic and pathologic processes. The discovery of a hypothalamic factor, gonadotropin releasing hormone (GnRH) that causes the release of the gonadotropins, luteinizing hormone (LH) and follicle stimulating hormone (FSH) from the pituitary gland has subsequently led to its synthesis, and use in clinical medicine. New knowledge regarding the control of gonadotropin secretion has thus emerged. There is increasing evidence that the seminiferous tubules elucidate a substance, inhibin, which selectively inhibits FSH secretion. The final identification of such a substance might have far-reaching implications in the regulation of male fertility. Added to all this is the rapid development of new technology which has allowed methodological advances not previously available.

In the human male, physiologic processes have been easier to study than have pathologic ones. Consequently, our knowledge of metabolic and endocrine factors that may be involved in male infertility or subfertility is still somewhat rudimentary. There are no good animal models that simulate human male infertility, and because only small amounts of tissue can be obtained from testicular biopsies, there is little human material available for study. Information obtained from light or electron microscopic examination of biopsy material may give morphologic data but very little indication of metabolic processes.

There have been many advances in our knowledge of biochemical, immunologic and morphologic parameters of human sperm; but this new knowledge has not brought us closer to a clearer understanding of pathological events in male infertility. Consequently, because we know so little about basic pathophysiological events, we have not yet been able to develop highly successful therapeutic regimens for human male oligo-azoospermia.

In this chapter, endocrine factors which may play a role in both male fertility and infertility will be discussed. In our laboratory, oligospermia is said to be present when the sperm count is less than 20×10^6/ml.

Hypothalamic-Pituitary Function: LH and FSH Response to GnRH

GnRH administered by a variety of routes causes the release of LH and FSH. LH in turn stimulates the secretion of testosterone by the Leydig cells while FSH promotes spermatogenesis after binding to Sertoli cells. The feedback inhibitory control of LH by testosterone is quite well known but the control of FSH secretion is less clearly understood.

The serum levels of LH and FSH in most normal men increase in a predictable fashion after the administration of GnRH. WOLLESEN et al. [27] studied time and dose responsiveness for the LH and FSH response to GnRH in normal males. They found that peak responsiveness occurred about 30 min after GnRH injection and that 100 μg of GnRH increased LH by 400–800% and FSH by 100–200% of baseline values. There have now been a large number of reports in which the gonadotropin increases after GnRH have been documented.

A few attempts have been made to determine whether there are differences in responses to GnRH among normospermic, oligospermic and azoospermic individuals. ISURUGI et al. [6] studied 11 young normospermic men, 7 with oligospermia (their definition of oligospermia was not given) and 7 with azoospermia. They found that normo- and oligospermic men had similar basal levels of LH, but azoospermic men had increased levels. The first two groups had the same LH increase (about 8 times baseline) after GnRH. Although LH increased only 4 times in the azoospermic group, the incremental increase was much higher than in the first two groups. Higher basal levels resulted in larger increments. FSH did not respond as briskly as LH. Basal levels of FSH were higher in the oligospermic group and higher yet in the azoospermic group. As with LH, the azoospermic group had a lower proportional rise of FSH but the incremental rise was higher than with the other two groups.

Table I. LH and FSH response to 100 μg GnRH administered i.v. to normospermic, oligospermic and azoospermic men

	a Mean LH, mIU/ml Time (min) after GnRH						
	0	0	20	30	45	60	120
Normospermic	4.4	5.3	34.2	36.5	32.7	27.13	13.8
Oligospermic	8.5	7.6	57.4	59.3	52.0	45.5	29.3
Azoospermic		10.2	78.6	77.3	68.7	61.7	45.6
	b Mean FSH, mIU/ml Time (min) after GnRH						
	0	0	20	30	45	60	120
Normospermic	3.2	3.8	6.5	6.9	6.7	6.8	6.6
Oligospermic	4.7	4.4	12.7	16.1	15.5	14.0	12.3
Azoospermic		9.4	33.3	38.7	37.2	36.3	33.4

MECKLENBURG and SHERINS [14] studied 6 normal men and 3 men with germinal aplasia. The azoospermic men had elevated basal levels of both LH and FSH and significantly increased responses to GnRH when compared to the normal groups, with FSH increases being greater than LH. NANKIN and TROEN [16] studied the LH and FSH response to GnRH in 8 subfertile men with sperm counts ranging from 30,000 to 36×10^6/ml; 7 men with sperm counts above 70×10^6/ml served as controls. Control LH values increased by an average of 170 ng/ml whereas in the subfertile group the mean LH increment was 441 ng/ml. FSH levels rose by an average of 62 ng/ml in the controls but by 225 ng/ml in the oligo-azoospermic group.

We have performed the GnRH test (100 μg in an i.v. bolus) on 4 azoospermic men, 6 with oligospermia (average sperm count less than 20×10^6/ml) and 5 with normospermia (average sperm count greater than 20×10^6/ml); LH and FSH were measured by the radioimmunoassay method of ODELL and SWERDLOFF [18].

The mean LH and FSH responses can be seen in table I. Although the number of men studied in each group is relatively small, a few important trends should be noted. Baseline LH levels in the azoospermic were significantly higher in the azoospermic group as compared to the normospermic group. Basal LH levels in the oligospermic group could not be distinguished from either of the other two groups. Although basal FSH levels in the azoospermic groups were not shown to be significantly different from basal FSH

Table II. Change in LH and FSH from baseline 30 min after GnRH

	Azoospermia	Oligospermia	Normospermia
LH, mIU/ml			
Mean baseline level	10.2	8.0	4.8
Mean 30-min level	77.3	59.3	36.5
Δ LH 30 min	67.1	51.3	31.7
FSH, mIU/ml			
Mean baseline level	9.4	4.5	3.5
Mean 30-min level	38.7	16.1	6.9
Δ FSH 30 min	29.3	11.6	3.4

of the other two groups, the tendency to increased levels is apparent. When a larger number of men were studied the difference did become statistically different (table III). From table II, it is seen that the incremental rise of both LH and FSH from baseline at the 30-min mark was highest in the azoospermic group and lowest for the men with normospermia. This is consistent with the results of others [13] demonstrating a hyperreactiveness to GnRH in primary gonadal failure. Our results suggest that Leydig cells in addition to seminiferous tubules have some type of defect in oligo-azoospermia accounting for the exaggerated LH and FSH response. In all three groups, LH was increased about 7-fold at 30 min; FSH was increased 4-fold in the two subfertile groups, whereas it only rose by a factor of 2 in the normospermic group.

Serum Levels of LH and FSH

Gonadotropin levels have been extensively studied over the years to determine whether a relationship between gonadotropin secretion and sperm production could be elicited. Early studies had to rely solely on bioassayable gonadotropins obtained from urinary extracts. Urine collections were often incorrect and bioassays were crude at best, hence meaningful data were hard to obtain. Finally, as sensitive and specific radioimmunoassay systems for both LH and FSH became available, much more intelligible information could be gathered and a better understanding of the relationship between LH and FSH secretion and testicular function emerged.

In 1970, LEONARD *et al.* [12] published an abstract describing increased serum FSH in certain patients with oligospermia. In 1972, ROSEN and WEIN-

Table III. Serum LH and FSH in 215 men grouped according to sperm count

Sperm count group $\times 10^6$/ml	Number	Mean LH, mIU/ml	Mean FSH, mIU/ml
>40	87	7.9	5.9
>30–40	14	7.2	4.0
>20–30	19	10.9[1]	7.0
>10–20	23	9.0	6.3
> 5–10	11	6.8	11.7[1]
> 0–5	29	12.2[1]	12.5[1]
0	32	15.4[1]	20.7[1]

[1] Different from LH or FSH at $>40 \times 10^6$/ml.

TRAUB [21] found that changes in sperm count could not be correlated with serum LH or testosterone levels, but they did find that FSH levels were elevated in patients with oligospermia or azoospermia, and that lower sperm counts were associated with higher FSH levels. Using a testicular biopsy score count method, JOHNSEN [7] determined the degree of spermatogenesis in 284 patients, all having primary testicular disorders. He correlated the biopsy score with total urinary gonadotropins and found that gonadotropins were elevated with late spermatogenic defects and did not rise further with earlier defects.

In further studies done by LEONARD *et al.* [11] a correlation between sperm count and plasma FSH levels could not be found. In their study the mean plasma FSH concentration of 60 oligospermic subjects was not significantly higher than that of 42 control subjects with normal spermatogenesis. However, their patients with azoospermia showed a consistent elevation of plasma FSH. They could discern no correlation between any of the germinal cell types and urinary FSH levels. WIELAND *et al.* [26] not only found an inverse relationship between sperm count and FSH in their oligospermic men but they also found elevated LH values which, however, did not correlate with sperm count. Serum testosterone levels in these men were normal.

There have been numerous investigations into the relationship between the gonadotropins and degree of spermatogenesis. Many of them suggest that seminiferous tubule insufficiency as evidenced by oligospermia or azoospermia may also be associated with a partial insufficiency of Leydig cell function; both of these dysfunctional states being manifested by elevations of LH and FSH. Our own studies support this contention. Table III indicates LH and FSH levels in 215 men with a wide spectrum of sperm counts. It can be seen that when compared to the group of men with sperm counts above 40×10^6/ml not only the azoospermic but also the oligospermic groups had significant elevations of LH and FSH. LH levels for the azoospermic and less than 5×10^6/ml group were statistically different from the group in which sperm count exceeded 40×10^6/ml. Surprisingly and without explanation, this was also true of the $20-30 \times 10^6$/ml group. FSH levels were also significantly higher when sperm counts of 10×10^6/ml or less were found.

These results suggest that men with oligospermia or azoospermia have total testicular insufficiency involving both major gonadal compartments, the seminiferous tubules and the Leydig cells, and that LH and FSH are secreted in increased concentrations in an attempt to stimulate the underfunctioning testes.

Testosterone and Dihydrotestosterone

It is the commonly held view that testosterone (T) concentrations in oligo-azoospermia are not different from those of normospermic men. Although most studies have reported serum T levels within the normal range regardless of sperm count, there are a number of reports which suggest that the secretion of T may also be mildly deficient when there is evidence of seminiferous tubular insufficiency. In men with germinal cell destruction and azoospermia secondary to chemotherapy for lymphoma [25] plasma T was within the normal range. However, the mean level in the men without germinal tissue was lower than in those with. This difference was not found to be statistically significant. NANKIN and TROEN [16] found the mean serum T in 12 normal men to be 559 ng/dl which was not significantly different from 462 ng/dl, the average of 26 oligospermic men. However, in the subgroup where sperm counts were less than 5×10^6/ml, the mean serum testosterone of 406 ng/dl was significantly below normal. In studying subfertile men, URRY et al. [24] found significantly decreased plasma T levels both in the group with azoospermia and with sperm counts less than 10×10^6/ml. They concluded that in male oligo-azoospermia not only were the seminiferous tubules insufficient but the Leydig cells, too, were functioning submaximally.

In our laboratory using radioimmunoassay techniques previously described [2, 19] we have measured the T and dihydrotestosterone (DHT) components of serum from men undergoing a fertility investigation. The results of these studies can be seen in table IV. Although the azoospermic

Table IV. Serum T and DHT related to different sperm count groups

Sperm count group $\times 10^6$/ml	Number	Mean T ng/dl	Mean DHT ng/dl
>20	76	525.4	58.3
>5–20	26	528.5	61.9
>0–5	24	506.6	60.2
0	18	449.9	60.5

Table V. Seminal plasma T and DHT concentrations in different sperm count groups

Sperm count group $\times 10^6$/ml	Number	Mean T ng/dl	Mean DHT ng/dl
>40	32	19.3	42.8[1]
>30–40	8	8.5	35.5
>20–30	7	11.9	30.0
>10–20	11	9.5	29.1
>0–10	15	10.9	21.5
0	4	8.3	13.1[2]

[1] Significantly different from all other groups (except >30–40).
[2] Significantly different from all other groups.

and severely oligospermic ($<5 \times 10^6$/ml) groups do have lower T levels, these are not significantly different from the normospermic group ($>20 \times 10^6$/ml). There were no differences in serum DHT.

Extensive studies on seminal plasma androgens have not been performed. We postulated that serum androgen concentrations may not accurately reflect events within the seminiferous tubule and that intratesticular levels of T and DHT as measured in seminal plasma may be more indicative of local conditions. Whether seminal plasma androgen concentrations in fact reflect intratesticular androgen metabolism remains an open question. Table V shows mean T concentrations within various sperm count groups. No differences could be elicited. Also shown in table V are the seminal plasma DHT values for each sperm count group. DHT levels increased with increasing sperm count, the highest concentrations being measured when sperm count exceeded 40×10^6/ml and the lowest concentrations being found in the azoospermic group.

An attempt was made to determine whether differences in sperm motility measured as the percent of actively forward moving sperm at 1 h could be

Table VI. Seminal plasma T and DHT concentrations correlated with sperm motility

Sperm motility group, %	Number	Mean T ng/dl	Mean DHT ng/dl
>40	41	10.6	37.3
>30–40	17	10.8	33.1
>20–30	14	9.5	35.7
>10–20	12	12.2	30.9
>0–10	10	9.8	27.4
0	9	11.4	18.6[1]

[1] Significantly different from all other groups.

related to variations in seminal plasma T or DHT. These results are shown in table VI. The only difference that could be detected was in the DHT level of the zero motility group which was significantly different from all other groups.

Several important observations can be made from the results of the above studies. One of these is the fact that DHT is the predominant androgen in seminal plasma. Furthermore, DHT and not T fluctuates according to both sperm count and sperm motility. Seminal plasma DHT is lowest when sperm count and motility are lowest. This suggests that abnormalities in intratesticular androgen concentrations may be reflected in decreased fertility as evidenced by either decreased sperm count, decreased sperm motility or both.

Estrogen and Prolactin

In 1969, KIAVOLA and JOHANSSON [8] reported that the mean urinary excretion of estrone was markedly lower in their pathospermic group of men than in the normospermic group. NANKIN and TROEN [16], on the other hand, found no significant relationship between serum estradiol and sperm count, although the mean level was higher than normal in their oligospermic group. In a small number of men studied in our laboratory we did find a relationship between serum estrogen levels and sperm count groups. Estrogens were measured by the method of ABRAHAM *et al.* [1]. Table VII demonstrates our results. With sperm counts less than 10×10^6/ml serum estrone levels were significantly elevated when compared to levels in normospermic men (sperm count $>20 \times 10^6$/ml). Estradiol measurements gave incongruous

Table VII. Serum estrogens correlated to sperm count group

Sperm count group × 10⁶/ml	Number	Mean estrone pg/ml	Mean estradiol pg/ml
>20	22	59	37
>10–20	6	55	39
> 0–10	14	64[1]	46[1]
> 0	7	65[1]	44

[1] Significantly different from >20.

Table VIII. Serum prolactin levels related to different sperm count groups

Sperm count group × 10⁶/ml	Number	Mean prolactin ng/ml
>40	21	10.4
>30–40	7	13.3
>20–30	4	9.4
>10–20	6	14.9
> 0–10	11	11.8
0	11	10.3

results in that the group whose sperm counts were between zero and 10×10^6/ml had significantly different values from those with normospermia but the azoospermic group did not. The small numbers studied and the large standard error may account for this discrepancy.

Because of an insufficient number of subjects studied it is difficult to draw any firm conclusions from these results. Because gonadotropins are elevated in oligo-azoospermia, it is not likely that an elevation of estrogens exert a suppressive influence on the hypothalamic-pituitary axis. Whether the elevated estrogens are a cause or an effect of inadequate spermatogenesis remains to be elucidated.

We recently became interested in the possible relationship between prolactin secretion and sperm production. SHETH et al. [23] found no significant correlation between serum prolactin levels and the androgenic and gametonic function of the testis. For our subjects serum prolactin was measured by the radioimmunoassay technique of HWANG et al. [5] at the Protein Hormone Laboratory of the Toronto General Hospital. We, too, found no correlation between prolactin and sperm count (table VIII).

Inhibin

The interrelationship of Leydig cell function in the testis and LH secretion from the pituitary gland is quite well established. LH binds to Leydig cells to cause the production and secretion of testosterone which in turn feeds back to the hypothalamic-pituitary axis to modulate further LH secretion [17]. The control of FSH secretion is not so clearly understood. It is known that FSH binds to Sertoli cells and probably participates in spermatogenesis via the elaboration of cyclic AMP [4]. But how events in the testes subsequently influence FSH secretion has not yet been definitively established. The current evidence, which is rapidly increasing, suggests that the seminiferous tubules elaborate a substance, inhibin, which suppresses the secretion of FSH by the pituitary gland.

For many years, the bulk of the evidence for the existence of inhibin has been found in clinical or experimental states in which seminiferous tubules were deficient and Leydig cells ostensibly remained intact. In these states, sperm production was reduced or absent, LH and testosterone concentrations were normal and FSH secretion was elevated. The implication was that when inhibin-producing elements were destroyed, there was no longer a damper on the elaboration of FSH which could now rise freely. Some of the human data already quoted in this chapter demonstrate that in azoospermia or in severe oligospermia in which there is some type of lesion within the seminiferous tubules serum, FSH levels rise, suggesting the loss of inhibin in these states. PAULSEN [20] found that when human volunteers had testicular irradiation, FSH levels rose and when sperm production returned to normal, FSH levels also became normal. This study suggested a radiation-induced defect of inhibin production which spontaneously resolved itself with return of seminiferous tubule function so that sperm production and inhibin secretion became normal.

In our laboratory we demonstrated that X-irradiation of rat testes was associated with cessation of spermatogenesis and rise of FSH [3]. LH did rise, but only about 7 weeks after irradiation. This was probably indicative of a delayed radiation effect to the Leydig cells. Testosterone secretion remained unchanged.

Several reports have now appeared indicating that purified extracts of various testicular components can selectively cause the suppression of FSH secretion without affecting LH. Materials which have been used have been rete testis fluid [22] or various testicular extracts [9, 15]. This has provided direct evidence for the existence of a specific testicular FSH suppressor.

DE KRETSER [10] has discussed some of the implications of inhibin. It becomes apparent that any substance which suppresses FSH secretion while leaving LH intact may have potential usefulness as a chemical contraceptive agent for men. Research along this line is currently in progress. Whether inhibin plays a role in male infertility remains to be determined. One could conceive of a state of primary hypersecretion of inhibin which would keep FSH secretion in check and hence inhibit spermatogenesis. This might explain the lack of FSH responsiveness after GnRH administration which is observed in some men. Since inhibin has not yet been clearly isolated, a highly sensitive assay system for it has not been developed, and some of these hypotheses cannot be tested.

Conclusions

Infertility or subfertility remain among the commonest and least understood medical problems of young men. Idiopathic oligospermia or azoospermia do not easily lend themselves to experimental study, and to date, the information that we have obtained is based not on our understanding of biological events within the seminiferous tubule but rather on less sensitive modalities such as hormone levels in various body tissues. From these studies it is only possible to make a number of inferences.

The evidence suggests that in both azoospermia and oligospermia, LH and FSH levels are elevated. Furthermore, there is an exaggerated gonadotropin response to LRH in both these conditions and serum testosterone tends to be lower at least in the azoospermic group. These gonadotropin changes and responses suggest that not only is the seminiferous tubule defective in male subfertility but that the Leydig cells, too, are not functioning normally. In other words, the entire testis in men whose sperm counts are significantly suppressed has some degree of insufficiency.

We have provided further evidence to show that in oligospermia and azoospermia androgen metabolism is abnormal in that DHT in seminal plasma is reduced when sperm counts are low or when sperm motility is absent. When sperm counts are very low, serum estrogen levels tend to be elevated.

Why so many men have relative testicular insufficiency is unclear. Until we can begin to understand the pathophysiological factors involved it is unlikely tpat a highly effective therapeutic regimen to enhance spermatogenesis will be found. Perhaps environmental factors such as air or radiation pollution are playing havoc with highly sensitive testicular tissue. Perhaps in some way the accelerated and anxiety-producing life style we lead is having an adverse influence on testicular function. Could it be that nature is sending us a message, a message that the world is getting too small for its ever increasing population?

Acknowledgements

The author wishes to acknowledge the technical assistance rendered by JEFFREY KEENE, MONIKA DUTHIE and ORA PALTIEL and the secretarial services of MORAG SMITH and ESTHER RAYNAI. The author expresses his gratitude to the Medical Research Council of Canada for financial assistance and to both the National Pituitary Agency of the NIAMDD and the Medical Research Council of Great Britain for reagents used in the radioimmunoassays of LH and FSH.

References

1. ABRAHAM, G.E.; HOPPER, K.; TULCHINSKY, D.; SWERDLOFF, R.S., and ODELL, W.D.: Simultaneous measurement of plasma progesterone, 17-hydroxyprogesterone and estradiol-17-β by radioimmunoassay. Analyt. Lett. *4:* 325 (1971).
2. BAIN, J.; GROVER, P.K.; SWERDLOFF, R.S., and ODELL, W.D.: One column chromatography and simultaneous radioimmunoassay (RIA) of testosterone (T) and dihydrotestosterone (DHT). Abstract No. 40. J. Steroid Biochem. *5:* 304 (1974).
3. BAIN, J. and KEENE, J.: Further evidence for inhibin: change in serum LH and FSH levels after X-irradiation of rat testes. J. Endocr. *66:* 279 (1975).
4. DORRINGTON, J.H. and FRITZ, I.B.: Effects of gonadotrophins on cyclic AMP production by isolated seminiferous tubule and interstitial cell preparations. Endocrinology *94:* 395 (1974).
5. HWANG, P.; GUYDA, H., and FRIESEN, H.: A radioimmunoassay for human prolactin. Proc. natn. Acad. Sci. USA *68:* 1902 (1971).
6. ISURUGI, K.; WAKABAYASHI, K.; FUKUTANI, K.; TAKAYASU, H.; TAMAOKI, B-I., and OKADA, M.: Responses of serum luteinizing hormone and follicle stimulating hormone levels to synthetic luteinizing hormone-releasing hormone (LH-RH) in various forms of testicular disorders. J. clin. Endocr. Metab. *37:* 533 (1973.)
7. JOHNSEN, S.G.: Investigations into the feedback mechanism between spermatogenesis and gonadotropin level in man; in ROSEMBERG and PAULSEN The human testis, p. 231 (Plenum Press, New York 1970).
8. KIAVOLA, S. and JOHANSSON, C.-J.: Excretion of gonadotropins (FSH and LH), 17-ketosteroids and oestrone in men with normal and abnormal spermatogenesis. Annls Chir. Gynaec. Fenn. *58:* 272 (1969).
9. KEOGH, E.J.; LEE, V.W.K.; RENNIE, G.C.; BURGER, H.G.; HUDSON, B., and KRETSER, D.M. DE: Selective suppression of FSH by testicular extracts. Endocrinology *98:* 997 (1976).
10. KRETSER, D.M. DE: The regulation of male fertility: the state of the art and future possibilities. Contraception *9:* 562 (1974).

11 LEONARD, J.M.; LEACH, R.B.; COUTURE, M., and PAULSEN, C.A.: Plasma and urinary follicle stimulating hormone levels in oligospermia. J. clin. Endocr. Metab. *34:* 209 (1972).
12 LEONARD, J.M.; LEACH, R.B., and PAULSEN, C.A.: Interrelationship of follicle-stimulating hormone (FSH) and spermatogenesis. Clin. Res. *28:* 169 (1970).
13 MARSHALL, J.C.: Investigative procedures. Clin. Endocr. Metab. *4:* 545 (1975).
14 MECKLENBURG, R.S. and SHERINS, R.J.: Gonadotropin response to luteinizing hormone-releasing hormone in men with germinal aplasia. J. clin. Endocr. Metab. *38:* 1005 (1974).
15 MOUDGAL, N.R.; NANDINI, S.G., and LIPNER, H.: Regulation of plasma FSH levels by ovine testicular extract (OTE). Programme Int. Symp. on Hypothalamus and Endocrine Functions, Quebec City 1975, Abstr. No. 33, p. 29.
16 NANKIN, H.R. and TROEN, P.: Endocrine profiles in idiopathic oligospermic men; in HAFEZ Human semen and fertility regulation in men, p. 730 (Mosby, St. Louis 1976).
17 ODELL, W.D. and MOYER, D.L.: Dynamic relationship of the testis to the whole man; in Physiology of reproduction, p. 82 (Mosby, St. Louis 1971).
18 ODELL, W.D. and SWERDLOFF, R.S.: Radioimmunoassay of luteinizing and follicle stimulating hormones in human serum; in HAYES, GOSWITZ and MURPHY Proceedings of the program of radioisotopes in medicine: *in vitro* studies. AEC Symposium Series No. 13, p. 165 (AEC, Oak Ridge 1968).
19 ODELL, W.D.; SWERDLOFF, R.S.; BAIN, J.; WOLLESEN, F., and GROVER, P.: The effect of sexual maturation on testicular response to LH stimulation of testosterone secretion in the intact rat. Endocrinology *95:* 1380 (1974).
20 PAULSEN, C.A.: In discussion; in ROSEMBERG Gonadotropins, p. 163 (Geron-X Inc., Los Altos 1968).
21 ROSEN, S.W. and WEINTRAUB, B.D.: Monotropic increase of serum FSH correlated with low sperm count in young men with idiopathic oligospermia and aspermia. J. clin. Endocr. Metab. *32:* 410 (1972).
22 SETCHELL, B.P. and JACKS, F.: Inhibin-like activity in rete testis fluid. J. Endocr. *62:* 675 (1974).
23 SHETH, A.R.; JOSHI, L.R.; MOODBIDRI, S.B., and RAO, S.S.: Serum prolactin levels in fertile and infertile men. Andrologie *5:* 297 (1973).
24 URRY, R.L.; DOUGHERTY, K.A., and COCKETT, A.T.K.: Correlation between follicle stimulating hormone, luteinizing hormone, testosterone and 5-hydroxyindole acetic acid with sperm cell concentration. J. Urol. *116:* 322 (1976).
25 VAN THIEL, D.H.; SHERINS, R.J.; MYERS, G.H., and DEVITA, V.T.: Evidence for a specific seminiferous tubular factor affecting follicle-stimulating hormone secretion in man. J. clin. Invest. *51:* 1009 (1972).
26 WIELAND, R.G.; ANSARI, A.J.; KLEIN, D.E.; DOSHI, N.S.; HALLSBERG, M.C., and CHEN, J.C.: Idiopathic oligospermia: control observations and response to cisclomiphene. Fert. Steril. *23:* 471 (1972).
27 WOLLESEN, F.; SWERDLOFF, R.S., and ODELL, W.D.: LH and FSH responses to luteinizing-releasing hormone in normal, adult, human males. Metabolism *25:* 845 (1976).

J. BAIN, BScPhm, MD, MSc, FRCP(C), Departments of Medicine and Obstetrics and Gynecology, University of Toronto and Reproductive Biology Unit, Mount Sinai Hospital, *Toronto, Ont.* (Canada)

Infection of the Male Reproductive Tract

B. Dahlberg

Department of Obstetrics and Gynecology, University of Lund, Malmö

Infertility was almost generally considered to be purely a female ailment in the 1950s. A decade later the conception that the male might be responsible for about 50% of the infertility, was accepted by doctors and most male subjects.

In the 1970s it is now recognized that male factors are causative in a significant number of infertile unions. Investigators during the last few years have shown that in many cases of male infertility the male might suffer from an infection in the reproductive system, sometimes without symptoms. Even a 'fertile' male may thus not always be fertile.

Some infections may lead to a chronic disease with irreversible sterility. Other infections may lead to short-term or long-term infertility. The consulting doctor's question will always be: 'How do I know if it is the male that is infertile? If he is, what is the cause of his infertility and how will I treat him?'

The routine investigation of an infertile marital unit comprises among other things a semen analysis which is usually examined at a special laboratory equipped for these examinations. If the answer from the laboratory says that the semen test is normal, and this will be the case with most laboratories using the prevalent methods, the doctor will gratefully accept this as a fact and tell the male patient that at least he is all right. If on the other hand the laboratory answers that the test is not normal or subnormal, it is up to the doctor to decide what to do. As the causes of male infertility become clearer and means of a correct diagnosis are established it is also evident that new parameters in the technique of semen analysis are important for the evaluation of the treatment and for the final and subsequent pregnancy in the female consort. Infertility in a marital unit is dependent on the interaction between male and female and treatment should be given to both.

However interesting and scientific a new method may be, there is still only one criterion of its success and that is a pregnancy, or if the percentage of pregnancies is high or higher than with other methods.

With recent findings that a high percentage of healthy males [6] in an infertile unit can be carriers of a silent infection in the reproductive tract, usually the prostate and vesicles, leading to asymptomatic bacteriospermia, treatment will now offer new hope for the marital unit.

Bacteria and Sperm Motility

Recent interest has been directed toward determining the significance of genital tract infections in the male and their effect on sperm motility and viability. Prostatic massage in normal, asymptomatic males has revealed wide variation in white cell levels in prostatic secretions [16]. Decreased motility and clumping of spermatozoa occurs immediately on mixing fresh ejaculate with *Escherichia coli* suspension in excess of 10 organisms/ml, whereas endotoxin has no effect on sperm [23]. There was no change in sperm motility in observations for 1 h with colony counts from 500 to 100,000 of *E. coli* [9]. Clumping occurred at concentrations of 10^7 and 10^8 with additional decrease in motility. The authors concluded that *E. coli* in significant concentrations specifically inhibits sperm motility and causes sperm clumping, thereby diminishing the amount of motile sperm available for conception.

Men with fertility problems associated with genital infections showed improvement of semen quality after the treatment of infection [19].

A chronic silent infection in the prostatic region can cause infertility in the male and an immune response in the female [4]. Examination of the male in cases of infertility should always include a culture of prostatic fluid by massage of the gland and of a sample of semen. This should be done after a period of continence, otherwise it is of little value.

Staphylococcus albus which is very often found in the urethra of the male and sometimes in cultures from semen samples is usually considered nonpathogenic. *In vitro* tests in our laboratory show that some strains of *S. albus* inhibit sperm motility. In a study of 190 infertile male subjects with infertility periods from 3 to 15 years it was shown that about 90% had various strains of bacteria in their semen. Normal controls were totally negative if they never had had intercourse. Among the normal fertile men half were negative and half had positive cultures of *S. albus* [6].

Asymptomatic Bacteriospermia

Genuine infertility in males is probably not as common as is believed. Infertility in a marital unit is dependent on the interaction between male and female. In most couples with infertility problems, the male seldom or never has any symptoms of disease or ailments. Furthermore the routine semen analyses from these males show very little or no deviation from normality. Therefore a 'normal' semen test in a male with infertility must be viewed with scepticism. As male infertility has been poorly understood, it is important to find factors that can help to diagnose the cause and means of treatment to restore fertility. New parameters in semen analyses are discussed below.

Examination of the male in cases of infertility should always comprise a culture of prostatic fluid obtained by massage and of seminal plasma.

The physiological ejaculation represents, beside a sperm specimen, also fluid from the prostate gland and the seminal vesicles. A microbial infection in this region can under given circumstances be discovered by bacteriological cultures of the semen specimen. The time for incubation is essential. A significant difference in the result showed that the optimal period was 5 days of abstinence. No significant difference has been observed in specimens taken between 5–6 and over 6 days.

In 190 men without any symptom of infection, between 22 and 52 years of age, with infertility periods between 3 and 15 years, semen cultures were taken [6]. Two groups of controls were used which consisted of group A, 10 young men under 18 years who maintained that they had not had intercourse previously, volunteered semen specimens, and group B, 10 males with proven fertility also delivered semen specimens. 290 bacteriological cultures were performed. In 33 consecutive cases evaluation of *Mycoplasma hominis* was done. All specimens were examined for *Trichomonas vaginalis* (table I).

Group A were all negative. Group B had 3 *T. vaginalis* and 5 *S. albus*. In the 190 infertiles 10 were negative. 90% had positive cultures with one or more strains. *Trichomonas* was observed in 89%. No *gonorrhea* was found.

Etiology

As most of the positive cultures found are normally represented in the colon or rectum it is obvious that contamination from the anal region will be common. In an interview of 6,000 women on personal habits of hygiene, e.g. after urination and bowel movement, 80% answered that they dried

Table I. Microorganisms found in sperm of infertile men and controls

Organisms	Number of cultures	%	20 Controls	
			10 precoital	10 fertile
Staphylococcus albus	94	49		5
Staphylococcus aureus	4			
Enterococci	77	41		
Streptococci Alpha	37	19		
B	1			
Anaerobes	4			
Coli	23	12		
Proteus	10			
Diphtheroid	10			
Klebsiella	6			
Pyocyaneus	3			
Pseudomonas	2			
Other	17			
Trichomonas	170	89		3
Candida	8			
Negative	10		10	5

The study included 190 men 22–52 years of age who had been infertile for periods ranging from 3 to 15 years. Of the infertile men 90% had positive cultures with one or more strains. Precoital controls all had negative cultures. Fertile controls: 5 were negative, 5 had *S. albus* and *Trichomonas*.

themselves starting in the anal area cleansing toward the vulval and urethral area. This habit is learned and inherited from mother to daughter. Only 12% cleansed properly from front to back to avoid contamination. Another cause of contamination is the widespread belief that during intercourse the penis will find its way in the vagina. In an interview with 500 couples, 90% answered that this was the case. Contact of the penis with the anus will contaminate both man and woman. Other sources of contamination are anal intercourse and orogenital intercourse.

The effect of the asymptomatic bacteriospermia is in most cases a reduced longevity of the spermatozoa. Figure 1 shows a case of a 28-year-old man with 4 years of infertility, the first sperm tests showed only 30% motile sperm with a longevity of only 2.5 h. After 6 months of treatment with antibiotics the sperm test showed 50% motile sperm with an increased longevity. Finally after 12 months treatment the test was normal and the wife became pregnant. A normal healthy child was born. Figure 2 shows the same pattern in 10 cases before and after treatment.

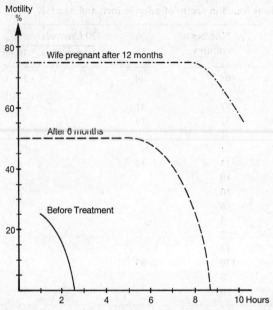

Fig. 1. Case C.N., 28 years, 4 years of infertility. Longevity and motility *in vitro* before treatment with antibiotics, after 6 months treatment and after 12 months treatment. When longevity and motility were normal, the wife became pregnant.

Another observed effect is that of agglutination which however does not appear in all cases. After antibiotic treatment agglutination disappeared. Agglutination reduces the number of motile sperm available for conception. Figure 3 shows case profiles.

Treatment

Long-term antibiotic treatment is often essential for a good result. Treatment regimen is as follows:

Trimethoprim (Wellcome) 160 mg twice a day for 4 months. After that the dose can be reduced to 80 mg twice a day if the culture is negative. *Pivampicillin* (Lövens) 350 mg 3 times daily continuously. *Erythromycin* (Abbott) 500 mg twice daily continuously. *Cephalexine* (Glaxo) 500 mg twice daily continuously. *Mycostatin* (Squibb) 500,000 IU 3 times daily for 2 weeks. The same dose is taken 1 week every month for 6 months.

Treatment goes on until the sperm test is normal or until the female partner is pregnant. If the sperm tests are very poor, treatment can take from 6 to 12 months. If the tests are subnormal treatment may be shorter. The

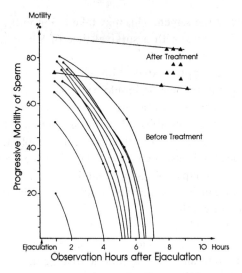

Fig. 2. Sperm tests in 10 infertile men with asymptomatic bacteriospermia. After 1 h observation, motility was normal in 9 of 10 cases; maximum longevity was 7 h. After treatment with antibiotics, antitrichomonal substances, and coitus with condoms for 6–12 months, all 10 sperm tests were normalized to 80%, with motility of more than 8 h. Wives of all 10 men became pregnant.

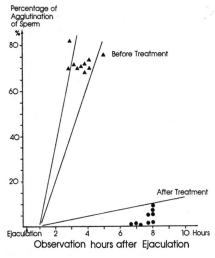

Fig. 3. Agglutination of spermatozoa of 10 men with asymptomatic bacteriospermia and infertility. Agglutination is not observed in the first hour. After 4 h, the majority of spermatozoa are clotted. After antibiotic and antitrichomonal treatment, agglutination diminished or disappeared. Agglutination of spermatozoa diminishes the number of free motile sperm.

first stage is a negative culture of the semen. This may take 1–3 months. The second stage is a prolonged longevity with a survival time of 8 h or more for over 50–60% of the spermatozoa.

The third stage is now an attempt at conception without danger of infection to the female. Treatment of the female should be done early. Usually a 2-week treatment with antibiotic and/or antitrichomonas substances is sufficient. As soon as a pregnancy is established, antibiotic treatment can be discontinued in the male. So as to avoid infections and complications during the pregnancy the male should use a condom. There is evidence that too short treatment may give a relapse of asymptomatic bacteriospermia which can affect the pregnancy causing miscarriage, cystitis and urinary infections and subsequent premature delivery.

Long-term treatment with antibiotics may sound hazardous with problems such as sensitization, overgrowth or lack of antibacterial effect, but these are not more frequent than in any other antibiotic treatment.

The seminal fluid should be examined after 2 months with a new culture and sensitivity test. Antitrichomonas and antimycotic treatment is given in repeated short-term intervals, for instance, 1 week per month. Ejaculation at least every 3–4 days is encouraged. In cases where the prostate consistency has not improved massage is recommended.

Results showed that with adequate treatment fertility was restored in 66% of 190 infertile men with asymptomatic bacteriospermia [6].

Asymptomatic bacteriospermia may lead to sexually transmitted diseases or ailments in the female consort which will account for part of the percentage of pregnancy failures or permanent sterility in the female, for instance tubal occlusion (10%), abortions (17%), urinary infections (17%), endometriosis (5%), immune response (25%) [6].

Prostatovesiculitis and Semen Viability

In males with decreased fertility there are high white cell counts in the semen [22]. This increase in white blood cells (WBC) is correlated with a corresponding increase in seminal fluid debris and a decrease in sperm motility. Infection of the genitourinary tract is associated with decreased fertility. QUESADA et al. [19] found a correlation between fertility problems and genital infections and improvement of semen quality was noted following the eradication of infection. Live bacteria have the ability to markedly reduce the motility and viability of human spermatozoa [9, 23].

Prostatic secretion, a major contributor to seminal fluid, normally contains 5–10 WBCs per high-powered field (WBC/HPF) and this number may increase several days post-intercourse [16]. However, in patients with chronic prostatitis, the WBC count can rise above 15/HPF, often higher, whereas sexual activity only transiently causes a rise in WBC count to 15-20/HPF. An increased semen WBC count exceeding 20 WBC/HPF can be used as an indication of genitourinary tract infection.

In cases with a semen WBC of 20/HPF, DERRICK and DAHLBERG [7] used metheonamine (Hiprex-Riker) and in treatment of over 100 patients achieved approximately a 30% response to treatment with pregnancies ensuing. There are three distinct responses to antibacterial therapy of prostatitis, as related to sperm motility. First a rapid increase in motility and a decrease in semen debris and WBC/HPF. This was evident in 2–3 weeks in the semen specimen and was a lasting result in some patients. The second response was quickly demonstrated by a decrease in the WBC/HPF and semen debris within 2–3 weeks, but with a rapid recurrence of debris and WBCs within 2–3 weeks after the medication has been discontinued. The third response showed little or no effect on the amount of WBCs and debris and sperm motility. The infection may have been caused by an organism, not responsive to Hiprex.

Male urethritis and prostatitis as diagnosed indirectly by debris and increased WBC count of above 20/HPF in fresh semen, may be associated with decreased fertility.

Genital Infections and Semen Viability

Reiter's Syndrome
BRODIE [3] was the first to describe 8 cases of a syndrome of urethritis, polyarthritis and inflammation of the eyes. REITER's description in 1916 of a case established the name for this sexually transmitted disease. Reiter's syndrome is described primarily as a bacterial urethritis, associated with bilateral conjunctivitis and polyarthritis, found almost exclusively in males. Suggestions have been made to implicate Chlamydia as the causative organism in Reiter's syndrome and other attempts have been made to implicate Mycoplasma but have not yet been proven. GORDAN *et al.* [14] found no evidence of this. There are no known effects of Reiter's urethritis on sperm viability. Tetracyclines or oxytetracyclines are administered.

Trichomonas vaginalis

T. vaginalis infestation in females is extremely common. Infestation in males ranges from 2.2 to 30.2% of urethritis cases [11]. Since the introduction of metronidazole (Flagyl), the incidence is much less. Symptoms are very similar to any other urethritis. *T. vaginalis* has been reported to cause hematospermia. The accepted methods of detecting *T. vaginalis* in the male urethra are a smear of any discharge, mixed with an equal amount of saline and observed under white light microscopy. Various staining methods are also helpful. There are some reports that *T. vaginalis* may interfere with sperm viability and motility [2, 20, 24]. *T. vaginalis* infestation only was not found to interfere with sperm viability [6].

The treatment for *T. vaginalis* is metronidazole (Flagyl) or tinidazole (Fasigyn) in one single dose [25].

Candida albicans

C. albicans is a fungus and is found in various places on the body of healthy human adults. The mouth, armpits, perineum, vagina and perirectal areas are most common sites. *C. albicans* lives in symbiosis with other organisms. *Lactobacillus*, usually found in the vagina and in the mouth and *E. coli* have been shown specifically to inhibit the growth of *Candida*. Quite frequently, during antibiotic therapy for other infections within the body, there will be a *Candida* overgrowth in the mouth and/or vagina and perineal areas. Current treatment is usually with a fungicidal agent, such as mycostatin (Squibb) 500,000 IU 3 times daily for 14 days with discontinuation of the systematic antibiotic.

C. albicans in significant colony counts, can definitely cause sperm clumping [7]. The sperm that are not affected by the clumping continue with normal motility. However, within 30 min after observation, approximately 50% of the sperm observed in the specimen had undergone clumping, thereby greatly reducing the effective motile sperm count in all specimens observed. In 190 cases of male infertility, we found 5% of *C. albicans* [6]. In 3 cases of infertility in the male where only *C. albicans* was found, we observed decrease of longevity. Treatment with mycostatin 500,000 IU 3 times daily for 1 month restored fertility.

Mycoplasma

DIENES and EDSALL [10] found 'L-organisms' or *Mycoplasma*. These organisms were isolated from excised Fallopian tubes with pelvic inflammatory disease. Subsequently, the mycoplasma organism has been discovered rather frequently in the genitourinary tract of both male and female and several reports indicate nongonorrheal urethritis can be caused by the *Mycoplasma* as often as in 40–60% of cases [1, 21]. *Mycoplasma* in the reproductive tract may be responsible for as many as 30–40% of cases of habitual abortion [12, 15, 17]. *M. hominis* has been isolated from one or both partners in 14.7% of 109 infertile couples [8]. On the other hand, DAHLBERG [6] found no positive cultures of *M. hominis* in 30 consecutive cases of infertile males where semen was cultured. *T. Mycoplasma* can be present in the male urethra and prostate without the subject being aware of its presence. It can also cause a typical urethritis with a thin, watery discharge. The organisms are extremely small and are not seen under usual white light microscopy and very special cultures are necessary for their isolation.

Mycoplasma only is probably not a cause of infertility, as it is mostly accompanied by positive cultures of other strains.

Treatment. Doxycycline 200 mg the 1st day and 100 mg daily subsequently is the drug of choice.

Chlamydia

Chlamydia trachomatis may be a more common cause of nongonorrhoic infection in both male and female than previously believed. In the male a postgonorrheal urethritis seems to be as common as the gonorrhoic urethritis. LYCKE *et al.* [18] found *C. trachomatis* to be an important etiologic agent in cases of acute salpingitis in female patients. The effect of *C. trachomatis* on the spermatozoon is unknown. As it may lead to a salpingitis in the female consort, it should also be treated as in the male. Doxycycline, trimethoprim and spectinomycin are effective *in vitro* [13].

Important Parameters in Semen Analyses

They are longevity, WBC, culture of bacteria, agglutination and pH.

Longevity

The fresh semen is examined as soon as possible usually within 1 h. The specimens are examined as to amount of living spermatozoa, percentage of progressive motility, longevity, WBC count, culture of bacteria, *T. vaginalis*, *C. albicans*, agglutination and percentage of abnormal cells. Among the most important criteria of infertility for a normal test is the amount of living spermatozoa per milliliter and the percentage of progressive motility. Unless longevity, i.e. the survival time of the spermatozoa, is observed, a very high percentage of semen tests will fall in the region of normality and the specimen will be considered normal even in cases of infertility.

To fertilize an ovum the spermatozoa must capacitate. The capacitation of the spermatozoa takes about 7 h. Thus the longevity of the spermatozoa is important. If the longevity is less than 7–8 h the spermatozoa are probably not able to fertilize.

Semen tests from normal fertile men show a significant difference in longevity when compared with semen tests from infertile men. Normal semen has a longevity of 8 h or more *in vitro* whereas semen from infertile men has less than 7 h of longevity [6].

For laboratory purposes longevity can be evaluated with repeated *in vitro* observations of the semen test.

The following method is used: In order to obtain reproducibility and accuracy a special pipette – the SMI micropettor (Genetec) – is used. The micropettor calibrated for 10 μl is used in two tests for each specimen and is rinsed in physiologic saline after each test. The drop is put on a slide and a thin glass put on top. Both specimens are put in a glass chamber with a filter paper humidified with saline to keep the specimens from drying and kept at room temperature. Drying immobilizes the spermatozoa and is lethal. The examination is started within 1 h and is repeated every third hour. The percentage of motile sperm is registered at each observation. Thus three observations can be done in an 8-hour period. The normal fertile semen tests all had a longevity of 8 h or more for 70% or more of the sperm. 7 h is the time for a spermatozoon to capacitate. An observation time of 8 h was therefore chosen as practical and feasible for laboratory work.

The cause of the reduced time of survival is mostly due to asymptomatic bacteriospermia.

With continuous treatment for several months the percentage of progressive motile sperm will augment as well the longevity. This evidently takes a long time and can be evaluated with the test described above. Usually at our laboratory, we have standardized examinations to every fifth month. In cases with low percentage of longevity, treatment can take up to 10–12 months.

WBC

The seminal plasma WBC count is usually raised in cases of prostatitis. A normal semen test shows few if any WBCs. WBC should always be reported with reference to the magnitude of magnification used so that a standardized procedure can be achieved and results will then be comparable to that of other laboratories.

A WBC with 20 or more per field and high power magnification of 400 HPF may be due to infection. Examination of WBC can sometimes be difficult in semen samples, as larger amounts of testicular cells may resemble WBCs. A Papanicolaou staining will in those cases easily show the difference and also give a well-colored spermatozoon.

Culture of Bacteria

It is recommended that seminal plasma culture should be done on all cases of infertility. However, it depends whether the bacteriological department responsible for the cultures is willing to do more than standard cultures. *Mycoplasma T* and *Chlamydia* are not evaluated in many laboratories. The routine methods for culture are, however, quite sufficient as most strains found in infertile males are quite 'common', i.e. those in the rectal and anal region. A sensitivity test for the right kind of antibiotic is also necessary. The chance of achieving a positive semen culture is enhanced if a 5- to 6-day continence period is observed. A semen test taken too soon after the last coitus will diminish the chance of getting a positive culture. Bearing this in mind we test for cultures on Fridays and not on Mondays. An instruction, written and oral, to the male is also necessary.

Agglutination

In many laboratories sperm agglutination is seldom observed as the clotting or clumping disappears if the specimens are shaken voluntarily or involuntarily. Agglutination or clumping of spermatozoa with either tail to tail or head to head or even mixed can easily be observed with the specimens in a humid chamber. The term agglutination is often reserved for the immunological test but the clotting and clumping of sperm are nothing else but agglutination and will thus be called so. The amount of clots per field can give a general impression of the nonprogressive sperm. An appreciation of the percentage of agglutinated sperm even if they are still motile but nonprogressive will be evidence of diminished fertility.

The higher the percentage of agglutinated cells the less free motile spermatozoa there will be available for fertilization of the ovum.

As far as the infertility is concerned there seems to be no difference between tail to tail and head to head agglutination.

pH

Seminal fluid normally has a pH of 6.5–7.0. With a pH of 7.5–8.0 a positive culture is usually found. It is a simple screening test to decide which samples should be cultured for bacteriospermia.

Summary

Investigations in the last decade have shown that the male factors can be causative in a significant number of infertile unions. Symptomless infection in the male reproductive tract, *asymptomatic bacteriospermia*, is common. The motility and longevity of the spermatozoa are reduced in such cases.

An aid to the diagnosis is an adequate method of taking the samples of prostatic fluid or seminal plasma and performing bacteriological cultures. Treatment is then started with the antibiotic indicated by the bacteriological department and continued till the sperm tests are normal. Results show that pregnancies can occur in about 66% of the cases after treatment.

References

1 AMBROSE, S.S., jr. and TAYLOR, W.W.: A study of the etiology, epidemiology and therapeusis of non-gonococcal urethritis. Am. J. Syph. Gonorrhea vener. Dis. *37:* 501 (1952).
2 ARGENZIANO, D.; DELUCA, M., and ROSSI, A.: Relationship between trichomonas and male infertility. Minerva derm. *42:* 388 (1967).
3 BRODIE, B.C.: Pathological and surgical observations on diseases of joints (Longman, London 1818).
4 DAHLBERG, B.: The lethal factor in infertility. Immunology in obstetrics and gynecology. Proc. 1st Int. Congress, Padua, p. 103 (Excerpta Medica, Amsterdam 1973).
5 DAHLBERG, B.: Proc. Human Semen and Fertility Regulation, Detroit (1975).
6 DAHLBERG, B.: Asymptomatic bacteriospermia as cause of infertility in men. Urology *8:* 563–566 (1976).
7 DERRICK, F,C.,jr. and DAHLBERG, B.: in HAFEZ Human semen and fertility regulation in men, p. 393 (Mosby, St. Louis 1976).
8 DELOUVOIS, J.; BLADES, M.; HARRISON, R.F.; HURLEY, R., and STANLEY, V.C.: Frequency of mycoplasma in fertile and infertile couples. Lancet *i:* 1073 (1974).
9 DEL PORTO, G. and DERRICK, F.C., jr.: Effect of bacteria on sperm motility. Urology *5:* 638 (1975).
10 DIENES, L. and EDSALL, G.: Observations on the L-organism of Klieneberger. Proc. Soc. exp. Biol. Med. *36:* 740 (1937).
11 GALLI, Z. and SYLVESTRE, I.: The present status of urogenital trichomonas, a general review article. Appl. Ther. *8:* 773 (1966).

12 GNARPE, H. and FRIBERG, J.: Mycoplasma and human reproductive failure. Am. J. Obstet. Gynec. *114:* 727 (1972).
13 GNARPE, H.: Sensitivity of clinical chlamydial isolates to various antimicrobials; in DANIELSSON, JUHLIN and MÅRDH, Genital infections. Wellcome Found. Symp., pp. 105–107 (Almquist & Wiksell, Stockholm 1975).
14 GORDAN, F.B.; AUAN, A.L.; STEINMAN, T.I., and PHILIP, R.N.: Chlamydial isolates from Reiter's syndrome. Br. J. vener. Dis. *49:* 376 (1973).
15 HORNE, H.W., jr.; KUNDSIN, R.B., and KOSASA, T.S.: The role of mycoplasma infection in human reproductive failure. Fert. Steril. *25:* 380 (1974).
16 JAMESON, R.M.: Sexual activity and the variations of the white cell count of the prostatic secretion. Investve Urol. *5:* 297 (1967).
17 KUNDSIN, R.B.: Mycoplasma in genitourinary tract infections and reproductive failure. Prog. Gynec. *5:* 275 (1970).
18 LYCKE, E.; JOHANNISSON, G., and EDMAR, B.: Chlamydia trachomatis infection and venereal disease. Proc. Swedish med. Soc. *16:* 275 (1976).
19 QUESADA, E.M.; DUKES, C.D.; DEEM, G.H., and FRENKLIN, R.R.: Genital infections and sperm agglutinating antibodies in infertile men. J. Urol. *99:* 106 (1968).
20 SCHMOR, J.: Sterility in urogenital trichomoniasis. Wien. med. Wschr. *120:* 808 (1970).
21 SHEPARD, M.C.: T-form colonies of pleuropneumonia-like organisms. J. Bact. *71:* 362 (1956).
22 SOKOL, S.; JACOBSON, C., and DERRICK, F.C., jr.: Methenadmine hippurate, a new use for an old drug. Male infertility. Urology *6:* 59 (1975).
23 TEAGUE, N.S.; BOYARSKY, S., and GLENN, J.F.: Interference of human spermatozoa motility by *E. coli.* Fert. Steril. *22:* 281 (1971).
24 WALTHER, H.: Trichomonas infection, clinical picture. Therapy and significance to fertility and sterility. Z. Haut-GeschlKrankh. *48:* 553 (1973).
25 WALLIN, J. and FORSGREN, A.: Tinidazole – a new preparation for *T. vaginalis* infections. Clinical evaluation of treatment with a single oral dose. Br. J. vener. Dis. *50:* 148 (1974).

BRUNO DAHLBERG, MD, Assistant Professor, Rudbecksgatan 80, *S-216 22 Malmö* (Sweden)

Immunological Aspects of Infertility[1]

Rudi Ansbacher

Department of Obstetrics and Gynecology, Brooke Army Medical Center, Fort Sam Houston, Tex.

For antibody production to occur, there must be an antigenic stimulus. The immune response that is initiated involves the processing of the antigen by the macrophage system with subsequent activation of lymphocytes. These lymphocytes proliferate and differentiate either to the thymic-independent humoral immune or the thymic-dependent cellular immune systems [5].

The B lymphocytes, which are not processed through the thymus gland, initiate the formation of immunoglobulins IgA, IgD, IgE, IgG, and IgM. The T lymphocytes, which are processed by the thymus gland, are instrumental in cellular immunity manifested by the delayed-skin reaction, graft-versus-host reaction, or homograft rejection [5]. T cells may also assist B cells in the formation of immunoglobulins [13].

Historical Aspects

The antigenicity of spermatozoa was first demonstrated by Landsteiner [29] and independently by Metchnikoff [33] in 1899. The use of semen to immunize the wife to prevent conception was introduced by Rosenfeld [41] in 1926 and repeated by Baskin [7] in 1932. The latter also utilized semen booster injections and serum testing. Mancini et al. [31] injected testicular homogenates mixed with Freund's adjuvant into men with prostatic cancer and demonstrated an 'allergic orchitis'. Present-day immunization of women

---

[1] The opinions or assertions contained herein are the private views of the author and are not to be construed as representing the views of the Department of the Army or the Department of Defense.

with spermatozoal fractions was attempted by OTANI et al. [37] but they reported anaphylactic reactions in 3 of 10 women [38].

Interest in immunologic infertility was stimulated by the independent reports of WILSON [50] and RÜMKE [42] in 1954 concerning the demonstration of sperm agglutinins in the serum and occasionally in the seminal plasma of men with unexplained sterility. FRANKLIN and DUKES [15] were the first to report a high incidence of sperm agglutinating activity in the serum of women with unexplained infertility. In 1969, FJÄLLBRANT [14] showed a reduction in the cervical mucus sperm penetration in couples with sperm antibodies.

Numerous investigators since the 1960s have tried to correlate sperm antibodies with infertility, but due to the nonstandardization of laboratory procedures and the disparate results reported from different patient populations without adequate control groups, it is difficult to evaluate the incidence of the phenomenon, the percentages reported, or the subsequent pregnancy rates, with or without treatment [3].

Theoretical Aspects

Female

Female sterility of immunologic origin involves mechanisms of isoimmunization. BEHRMAN [8] hypothesized that the absorption of spermatozoa by the mucosa of the upper vagina or the cervix or by the lower uterine endometrial glands, with phagocytosis and transport to the regional lymph nodes, might initiate antisperm antibody production. The primary response is probably mediated by IgM and the secondary response by IgG.

Why the majority of women do not form antisperm antibodies despite repeated exposure to spermatozoa is unknown, but it appears that unless a hyperimmune state is induced, the sperm antigen-antibody reaction does not develop, which could be protective to the woman. (IgG sperm antibodies are either agglutinating, immobilizing, or cytotoxic; at first, they are not too avid and probably are of the agglutinating type. Later, they become more avid, probably immobilizing or cytotoxic, i.e., hyperimmune.) In addition, there might be an incomplete response to the antigen (sperm) or blocking antibody might hinder the interaction of the antibody with the antigen.

BEHRMAN [8] also suggested that local or transvaginal immunization may cause the local production of antibodies that might have a considerable effect on fertility, but which might or might not produce enough antibodies to be detected in the serum. Agglutination or immobilization of spermatozoa might occur in the vaginal or cervical secretions in sufficient degree to cause infertility in a given couple, but in another couple, the concentration of spermatozoa might not be high enough to compensate for the antibody that is present (prozone effect).

Effects in the female include: (1) inhibition of sperm-cervical mucus interaction; (2) lower subsequent fertility rate and a higher abortion rate in women with antisperm antibodies compared to those without [23].

Male

Male sterility of immunologic origin may occur as a result of trauma to the testis, infection (subclinical, epididymitis, or orchitis), or blockage of the excretory ducts. Extravasation of spermatozoa into the interstitium, lymph vessels, or blood capillaries of the epididymis with subsequent transfer into the regional lymph nodes might initiate sperm antibody production [43]. Intraluminal phagocytosis plays a role in the disposal of dead spermatozoa in patients with obstructive azoospermia or with ligated vasa deferentia [39].

Another plausible theory is that there is a loss of tolerance to organ-specific antigens, but what mechanism might lead to the loss of tolerance? The testis appears to fulfill the criteria for a loss of tolerance because [5]:

(1) The male reproductive tract does not attain functional maturity until puberty. Therefore, no antigen is present when tolerance to self-antigens is developed.

(2) Spermatozoa have a conduit, the vas deferens, for egress from the body. Therefore, the individual does not recognize spermatozoa as antigens.

(3) A blood-testis barrier prevents the passage of serum proteins into the seminiferous tubules [22].

Effects in the male include: (1) interference with normal spermatogenesis; (2) direct effects on the spermatozoa which prevent fertilization of the ovum or cause defects in the zygote; (3) sperm transport changes.

Sperm Antigens and Sperm Antibodies

MENGE and FULLER [32] demonstrated 14 substances in the human ejaculate which were antigenic. Ten were testis specific in man and cross-reacted with chimpanzee, rhesus monkey, and baboon spermatozoa. Most probably coat the spermatozoa and their importance is still debatable.

TUNG [49] reported sperm antibodies to specific sites on the spermatozoa using immunofluorescent techniques (table I). TUNG [49] noted that

Table I. Sperm-specific antibody sites demonstrated by immunofluorescent techniques according to Tung [49]

Antigen	Immunoglobulin	Type
AC-1 (smooth fluorescence)	IgM, IgG, or both	'natural'
AC-2 (speckled fluorescence)	IgM and IgG	
Equitorial region		'natural'
Postacrosomal region		'natural'
Main piece of the tail		
Nuclear (fixes complement)	IgG	
Midpiece		'natural'

the 'natural' antibodies were present even in children, whereas the other three could only be demonstrated in men who had had bilateral vas ligations.

Antibodies have also been demonstrated against human sperm hyaluronidase, which are tissue and species specific [34]; human protamine I and II [45]; LDHx, which can be an autoantigen or isoantigen and which is sperm specific but not species specific [40]; and 'immobilizing antigen', which is a sperm surface (head and tail plasma membrane) membrane glycoprotein [35].

Site of Action

The cervix would appear to be one of the likely sites of action for the sperm antigen-antibody reaction. A reduction in cervical mucus penetration by spermatozoa treated with antispermatozoal antibodies from either rabbit or man was shown by Fjällbrant [14] using the capillary tube cervical mucus sperm penetration test [26]. He suggested that this was the likely mechanism involved in the infertility of couples with sperm antibodies.

IgA and IgG antibodies were found in samples taken from human endometrium, Fallopian tubes, cervix, and vagina [30]. In addition, the biosynthesis of immunoglobulins IgA and IgG was demonstrated in the cervical and uterine glandular tissues of rabbits [9] and women [10].

In 12 of 50 (24%) infertile females without an apparent infertility factor Hutcheson et al. [19] found an increase in plasma cells, which contained IgA, in their cervical mucus. Venereal disease researchers [24, 36] at the Center for Disease Control in Atlanta have shown that both males and females produce specific secretory IgA antibodies against the gonococcus in their genital tracts.

Sperm antibodies can be found in the cervical mucus extracts of infertile women as compared to normal controls; they were found in the IgG and IgA immunoglobulin classes, but not the IgM [47]. Interestingly, SHULMAN reported at the April, 1976 annual meeting of the American Fertility Society that 30% of his patients with positive sperm antibody activity in their cervical mucus had negative sera.

From the above, it appears that a local immune response to spermatozoal antigens can occur in the cervix, which may be unrelated to systemic sperm antibody production. If these immune antibodies in the cervical secretions can react with specific antigens, immobilization or agglutination of spermatozoa could occur during their transport from the cervical canal to the uterine cavity. Therefore, cervical mucus sperm interaction must be evaluated.

Cellular immunity, mediated by T lymphocytes, is finally receiving more attention by investigators, but to date there is little to report concerning an association with immunologic infertility.

Laboratory Tests

The commonly utilized laboratory procedures to test for the presence of sperm antibodies are listed in table II.

FRIBERG [16] has also utilized the tray agglutination test and noted that sera positive with the tube slide agglutination method have head-to-head sperm agglutination whereas those positive with the gelatin agglutination method show head-to-tail sperm agglutination.

Table II. Laboratory tests

Name of test	Material used	Activity fraction	References
Tube slide agglutination	serum, sperm	β-globulin	12, 15, 16
Gelatin agglutination	serum, sperm	IgG	12, 16, 25
Sperm immobilization	serum, sperm, complement	IgG and IgM	20, 21
Cytotoxic	serum, sperm		17
Immunofluorescence	serum, sperm		49
Capillary tube cervical mucus sperm penetration	cervical mucus, sperm		26
Sperm-cervical mucus contact	cervical mucus, sperm		27

Incidence

The probable incidence of sperm antibodies causing infertility is between 12 and 20% of those couples without any other explanation for their infertility. (Theoretically, if one sees 100 infertile couples, one can usually find an explanation for their infertility in 60. Of the remaining, 5–8 couples may have an immunologic basis for the infertility. This is based upon the author's unpublished data.)

Proposed Evaluation

A properly timed SIMS [48] – HUHNER [18] postcoital examination performed 24–36 h prior to expected ovulation, after 2 days of coital abstinence, and within 2 h of coitus, with the finding of immotile or agglutinated spermatozoa, with poor or no penetration of the cervical mucus by spermatozoa, should alert the clinician to the possibility that sperm antibodies are present [1–4, 6].

In vitro sperm penetration tests, such as the ones described by KREMER [26] and KREMER and JAGER [27] should be performed so that cross-matching can be accomplished to determine the significance of the antibody present and which partner carries it (wife's cervical mucus vs. husband's spermatozoa, donor's cervical mucus vs. husband's spermatozoa, wife's cervical mucus vs. donor's spermatozoa, and donor's cervical mucus vs. donor's spermatozoa) [6].

The sperm-cervical mucus contact test [27] has been used to differentiate immunologic and nonimmunologic agglutination. In a positive test the motile spermatozoa with forward progression become stationary with a shaking type of motility pattern as soon as contact between sperm and cervical mucus occurs, indicating the presence of sperm agglutinins.

Serum samples from both the wife and the husband should be forwarded to a laboratory specializing in sperm antibody tests prior to the initiation of therapy.

Therapy

Treatment of patients, male or female, with circulating antisperm antibodies is poorly defined. Continuous condom therapy has been advocated for sensitized women to prevent the repeated absorption of spermatic antigens [8, 11, 15].

A number of treatment modalities require further evaluation in properly controlled clinical trials. Examples include the inhibition of male autosensitization by testosterone suppression of spermatogenesis [44, 46]; the use of ascorbic acid, a potent reducing agent to overcome male autoagglutination [28]; and immunosuppressive agents, such as cortisone and its derivatives, to prevent or suppress the immune response in either male or female patients.

Artificial insemination with donor semen for the wives of husbands with antisperm antibodies may be indicated for some couples. Intrauterine insemination with 0.1–0.2 ml husband's semen or washed spermatozoa to bypass the cervix is another possibility. Adoption should be considered either when both partners are affected or when the wife has a sperm antibody titer that does not fall in response to condom therapy [3].

Implications

Both male autoimmunization and female isoimmunization seem to lead to a reduction in fertility not necessarily accompanied by refractory sterility.

Identification of specific sperm antigens, sperm antibodies, and sperm antigen-antibody binding sites is now being studied. More intensive investigation of sperm-cervical mucus interaction must be undertaken, especially to evaluate the immunoglobulins associated with sperm antibodies in the cervical mucus, the role of complement in sperm immobilization studies, and the possible cellular immune aspects of immunologic infertility.

Newer techniques are being developed to aid in the above. From these, clinical laboratory studies should be developed to enable clinicians to properly identify both women and men with reproductive immunologic abnormalities. Until such studies are available and universally applied, the disparities in results reported from various centers will continue.

References

1 ANSBACHER, R.; MANARANG-PANGAN, S., and SRIVANNABOON, S.: Sperm antibodies in infertile couples. Fert. Steril. 22: 298–302 (1971).
2 ANSBACHER, R.; KEUNG-YEUNG, K., and BEHRMAN, S.J.: Clinical significance of sperm antibodies in infertile couples. Fert. Steril. 24: 305–308 (1973).

3   ANSBACHER, R.: Sperm antibodies in infertile couples. Contemp. Obstet. Gynec. 2: 79–81 (December 1973).
4   ANSBACHER, R.: Significance of sperm antibodies in infertile couples. J. reprod. Med. 13: 48–50 (1974).
5   ANSBACHER, R.: Autoimmunity of spermatozoa; in Human semen and fertility regulation in men, pp. 265–267 (Mosby, St. Louis 1976).
6   ANSBACHER, R.: Cervical mucus: its possible role in immunologic infertility. Contemp. Obstet. Gynec. 8: 25–29 (October 1976).
7   BASKIN, M.J.: Temporary sterilization by injection of human spermatozoa, preliminary report. Am. J. Obstet. Gynec. 24: 892–897 (1932).
8   BEHRMAN, S.J.: The immune response and infertility: experimental evidence; in Progress in infertility, pp. 675–699 (Little, Brown, Boston 1968).
9   BEHRMAN, S.J.; LIEBERMAN, M.; UCHIYAMA, N., and ANSBACHER, R.: Immunoglobulin synthesis of the rabbit reproductive tract in vitro; in Pathways to conception: the role of the cervix and the oviduct in reproduction, pp. 237–244 (Thomas, Springfield 1971).
10 BEHRMAN, S.J. and LIEBERMAN, M.E.: Biosynthesis of immunoglobulins by the human cervix; in Biology of the cervix, pp. 235–249 (University of Chicago Press, Chicago 1973).
11 BEHRMAN, S.J.: The immune response and infertility; in Progress in infertility; 2nd ed., pp. 793–815 (Little, Brown, Boston 1975).
12 BOETTCHER, B.; KAY, D.J.; RÜMKE, P., and WRIGHT, L.E.: Human sera containing immunoglobulin and nonimmunoglobulin. Biol. Reprod. 5: 236–245 (1971).
13 COHEN, I.R. and WEKERLE, H.: Autosensitization of lymphocytes against thymus reticulum cells. Science 176: 1324–1325 (1972).
14 FJÄLLBRANT, B.: Cervical mucus penetration by human spermatozoa treated with anti-spermatozoal antibodies from rabbit and man. Acta obstet. gynec. scand. 48: 71–84 (1969).
15 FRANKLIN, R.R. and DUKES, C.D.: Antispermatozoal activity and unexplained infertility. Am. J. Obstet. Gynec. 89: 6–9 (1964).
16 FRIBERG, J.: Clinical and immunological studies on sperm-agglutinating antibodies in serum and seminal fluid. Acta obstet. gynec. scand., Suppl. 36, pp. 1–19 (1974).
17 HAMERLYNCK, J. and RÜMKE, P.: A test for the detection of cytotoxic antibodies to spermatozoa in man. J. Reprod. Fert. 17: 191–194 (1968).
18 HUHNER, M.: Sterility in the male and female and its treatment (Robman, New York 1913).
19 HUTCHESON, R.B.; ANDERSON, T.D., and HOLBOROW, E.J.: Cervical plasma cell population in infertile patients. Br. med. J. iii: 783–784 (1974).
20 ISOJIMA, S.; LI, T.S., and ASHITAKA, Y.: Immunologic analysis of sperm-immobilizing factor found in sera of women with unexplained sterility. Am. J. Obstet. Gynec. 101: 677–683 (1968).
21 ISOJIMA, S.; TSUCHIYA, K.; KOYAMA, K.; TANAKA, C.; NAKA, O., and ADACHI, H.: Further studies on sperm-immobilizing antibody found in sera of unexplained cases of sterility in women. Am. J. Obstet. Gynec. 112: 199–207 (1972).
22 JOHNSON, M.H.: The distribution of immunoglobulin and spermatozoal autoantigen in the genital tract of the male guinea pig: its relationship to autoallergic orchitis. Fert. Steril. 23: 383–392 (1972).

23 JONES, W.R.: The use of antibodies developed by infertile women to identify relevant antigens. Acta endocr., Copenh. *78:* suppl. 144, pp. 376–404 (1975).
24 KEARNS, D.H.; O'REILLY, R.J.; LEE, L., and WELCH, B.D.: Secretory IgA antibodies in the urethral exudate of men with uncomplicated urethritis due to *Neisseria gonorrhoeae.* J. infect. Dis. *127:* 99–101 (1973).
25 KIBRICK, S.; BELDING, D.L., and MERRILL, B.: Methods for the detection of antibodies against mammalian spermatozoa. II. A gelatin agglutination test. Fert. Steril. *3:* 430–438 (1952).
26 KREMER, J.: A simple sperm penetration test. Int. J. Fert. *10:* 209–215 (1965).
27 KREMER, J. and JAGER, S.: The sperm-cervical mucus contact test, a preliminary report. Fert. Steril. *27:* 335–340 (1976).
28 KUPPERMAN, H.S. and EPSTEIN, J.A.: Endocrine therapy of sterility. Am. Practnr Dig. Treat. *9:* 547–563 (1958).
29 LANDSTEINER, K.: Zur Kenntnis der spezifisch auf Blutkörperchen wirkenden Sera. Zentbl. Bakt. (Orig.) *25:* 546–549 (1899).
30 LIPPES, J.; OGRA, S.; TOMASI, T.B., and TOURVILLE, D.R.: Immunohistological localization of γG, γA, γM, secretory piece and lactoferrin in the human female genital tract. Contraception *1:* 163–183 (1970).
31 MANCINI, R.E.; ANDRADA, J.A.; SARACENI, D.; BACHMANN, A.E.; LAVIERE, J.C., and NEMIROVSKY, M.: Immunological and testicular response in man sensitized with human testicular homogenate. J. clin. Endocr. Metab. *25:* 859–875 (1965).
32 MENGE, A.C. and FULLER, B.: Testis antigens of man and some other primates. Fert. Steril. *26:* 473–479 (1975).
33 METCHNIKOFF, E.: Etudes sur la résorption des cellules. Annls Inst. Pasteur, Paris *13:* 737–769 (1899).
34 O'RAND, M.G. and METZ, C.B.: Tests for rabbit sperm surface iron-binding protein and hyaluronidase using the 'exchange agglutination' reaction. Biol. Reprod. *11:* 326–334 (1974).
35 O'RAND, M.G. and METZ, C.B.: Isolation of an 'immobilizing antigen' from rabbit sperm membranes. Biol. Reprod. *14:* 586–598 (1976).
36 O'REILLY, R.J.; LEE, L., and WELCH, B.G.: Secretory IgA antibody responses to *Neisseria gonorrhoeae* in the genital secretions of infected females. J. infect. Dis. *133:* 113–125 (1976).
37 OTANI, Y.; INO, H., and KAGAMI, T.: Antigenicity of human semen, sperm, and testis. Int. J. Fert. *10:* 143–149 (1965).
38 OTANI, Y.; INO, H.; INOUE, S., *et al.:* Immunization of human female with human sperm and semen. Int. J. Fert. *16:* 19–23 (1971).
39 PHADKE, A.M.: Fate of spermatozoa in cases of obstructive azoospermia and after ligation of vas deferens in man. J. Reprod. Fert. *7:* 1–12 (1964).
40 PRASAD, R.; MUMFORD, D., and GORDON, H.: Lactate and malate dehydrogenase and α-esterases in oligospermia. Fert. Steril. *27:* 832–835 (1976).
41 ROSENFELD, S.S.: Semen injections with serologic studies, a preliminary report. Am. J. Obstet. Gynec. *12:* 385–388 (1926).
42 RÜMKE, P.: The presence of sperm antibodies in the serum of two patients with oligozoospermia. Vox Sang. *4:* 135 (1954).
43 RÜMKE, P. and HELLINGA, G.: Autoantibodies against spermatozoa in sterile men. Am. J. clin. Path. *32:* 357–363 (1959).

44 RÜMKE, P.; AMSTEL, N. VAN; MESSER, E.N., and BEZEMER, P.D.: Prognosis of men with auto-spermagglutinins in the serum and the unsuccessful treatment with testosterone; in 2nd Int. Symp. on Immunology of Reproduction (Bulgarian Academy of Science Press, Sofia 1973).
45 RÜMKE, P.: Personal commun. (1975).
46 SCHOYSMAN, R.: Communication at the 6th Wld Cong. on Fertility and Sterility, Tel Aviv 1968 (unpublished).
47 SHULMAN, S. and FRIEDMAN, M.R.: Antibodies to spermatozoa. V. Antibody activity in human cervical mucus. Am. J. Obstet. Gynec. *122:* 101–105 (1975).
48 SIMS, J.M.: Clinical notes on uterine surgery with special reference to the sterile conditions (Wood, New York 1866).
49 TUNG, K.S.K.: Human sperm antigens and antisperm antibodies. I. Studies on vasectomy patients. Clin. exp. Immunol. *20:* 93–104 (1975).
50 WILSON, L.: Sperm agglutinins in human semen and blood. Proc. Soc. exp. Biol. Med. *85:* 652–655 (1954).

R. ANSBACHER, MD, MS, COL, MC, USA, FACOG, Department of Obstetrics and Gynecology, Brooke Army Medical Center, *Fort Sam Houston, TX 78234* (USA)

Genetic Aspects of Male Infertility

M.H.K. Shokeir

Division of Medical Genetics, University of Saskatchewan, Saskatoon, Sask.

Introduction

The attainment of fertility by the male depends on a concerted sequence of developmental events based on appropriate chromosomal make-up and genetic program. Differentiation steps guided by the information encoded in such a program and coordinated with parallel development of associated systems ensure normal morphogenesis of the genital system. Its maturation is further controlled by the genetic blueprint through neural and hormonal stimulation and sustained by a favorable internal and external milieu.

Infertility may result from alteration of the normal chromosomal constitution, both of the sex chromosomes and the autosomes; genetic errors; failure of normal differentiation of the gonads and the genitalia, both internal and external; teratogenetic malformations; arrest or deviation of the maturation processes; and interference with the subtle humoral and neural control. Furthermore, the establishment of the normal male gender and masculine self-image are themselves physiologically and psychosocially intricate events which are prone to potential abnormalities. Finally, heterosexual expression is a mandatory prelude to reproduction, and departure from it is probably, in some instances, genetically determined.

In this chapter no attempt will be made to enumerate exhaustively genetic causes of male infertility nor the major works published in this burgeoning field. Instead only the broad outlines will be depicted.

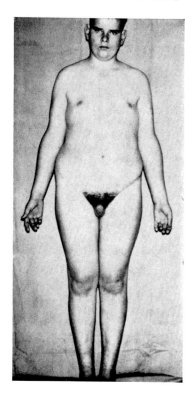

Fig. 1. A 30-year-old with Klinefelter syndrome showing gynecomastia and obesity.

Genetic Causes of Infertility

Chromosomal Causes

Disorders of the Sex Chromosomes

Whereas the Y chromosome determines maleness, it is the absence of the second X (or the presence of a single X) chromosome which permits the male to develop normal secondary sexual characteristics and be fertile. Departure from the XY sex chromosomal complement in the male usually leads to disorders of masculinization and infertility.

Anomalies of the X Chromosome

Klinefelter syndrome (fig. 1). The presence of more than a single X chromosome leads to a characteristic phenotypic abnormality which in the

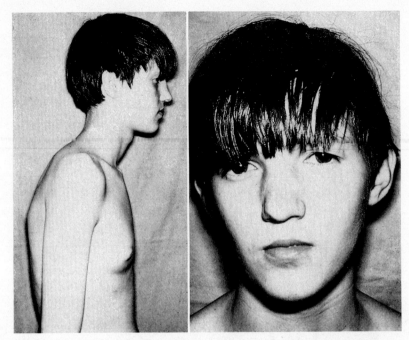

Fig. 2. A 18-year-old Klinefelter male with gynecomastia and absence of facial hair.

more frequent XXY complement is known as Klinefelter syndrome [43]. It is now recognized to be the most common single cause of infertility when associated with panhypogonadism [62].

Frequency: About 1 in every 500 males is affected.

Clinical features: The abnormalities which are not invariable include those of linear growth with tendency to increased stature. The ratio of upper to lower segments of the body may be low from childhood onwards because of the relatively long limbs. If untreated adults tend to be obese.

The hypogonadism dates from childhood with relatively small penis and testes, usually less than 2.5 cm each in length in the adult. Their consistency often becomes firm as age advances, in contrast with other forms of male hypogonadism wherein the testicular consistency is soft.

Infertility is the rule. Hyalinization and fibrosis of the seminiferous tubules under the influence of excessive gonadotrophin account for failure of spermatogenesis. Hormone production by the testes is likewise impaired. The average serum testosterone in adults with Klinefelter syndrome is around 50% that of the normal mean. This leads to inadequate virilization with gynecomastia occurring in 40%, diminished total muscle mass, gracile long limbs, delayed fusion of the epiphyses with continued linear growth, tendency to osteoporosis, poor facial and body hair growth (fig. 2), reduced libido and weaker potency, as well as behavior disorders. The latter tend to include immaturity, effeminacy, poor judgement, reticence or paradoxically unrealistic boastfulness and self-assertion. Homosexuality is more frequent and exclusive than in XY males [17].

Intelligence may be mildly impaired in these patients with the average IQ around 85 or 1 standard deviation below the mean of unaffected males [54].

Diagnosis: Dull mentality, behavior problems, small penis and testes with relatively tall stature may suggest the diagnosis during childhood. Subsequently, gynecomastia, inadequate virilization, small and especially firm testes, azoospermia and diminished libido, and worsening behavior problems make the diagnosis more clinically tenable.

The diagnosis may be confirmed by buccal smear examination. In the normal XY male it is negative for X-chromatin (Barr) body, whereas in Klinefelter males a proportion of the nuclei equivalent to that seen in females (about 25%) is X-chromatin positive (fig. 3a). The X-chromatin represents the genetically inactivated X chromosome. Full chromosome studies should follow in those males who are found to be X-chromatin positive or who show strong clinical evidence of the syndrome. At least 30 metaphase plates should be examined before the diagnosis or mosaicism can be excluded.

Etiology: The classic Klinefelter syndrome is due to the presence of an extra X chromosome giving an XXY sex chromosomal complement with a total chromosome count of 47 (fig. 3b) instead of the normal number of 46. The extra X chromosome may have been included either at the first meiotic division during spermatogenesis in the father or at either the first or second meiotic divisions during oogenesis in the mother. Should it have been the paternal spermatogenesis which was at fault it would have resulted in a sperm bearing both the X and Y chromosomes with a chromosome number of 24 instead of the normal haploid number of 23. On the other hand were it the maternal oogenesis which was affected, the nucleus of the ovum would have contained two X chromosomes with a total count of 24 chromosomes. It is fertilization of such an egg by a normal Y-bearing sperm or the fertilization of a normal egg by an XY-bearing sperm that results in an XXY (Klinefelter) zygote. The mechanism whereby such an error occurs is that of nondisjunction; when the two members of a pair of homologous chromosomes, such as the two X chromosomes in the female, fail to segregate to the opposite poles of the cell during the first meiotic division or the two sister chromosomes during the second meiotic division. Increased maternal age appears to be the most important single factor in the predisposition to chromosomal non-disjunction during meiosis [44]. Increased paternal age has no etiologic significance. The occurrence of a previous non-disjunction as shown in a trisomic sibling and familial clustering are two other possible contributory factors. Alternatively, non-disjunction may take place during mitosis of the newly developing XY zygote. The two X chromosomes that result from chromosomal replication during the first cell division of the fertilized egg may by accident migrate to one and the same pole resulting in a nucleus containing XXY sex chromosomal complement and another cell having only the Y sex chromosome. The latter will perish leaving a zygote which is uniformly XXY. Should the error occur at an early, but not the first, mitotic division a mosaic XY/XXY zygote will result. The phenotypic features in the child, adolescent and adult will therefore be intermediate between normal and XXY Klinefelter males. Such individuals have a better potential prognosis for testicular function including possible spermatogenesis with consequent fertility. Although 22% of X-chromatin positive male newborns were found on intensive chromosome studies to be mosaic XY/XXY, only a small proportion of these showed evidence of potential fertility.

Management: Testosterone replacement therapy is the mainstay of treatment of affected males [12] although it is of no help in reversing their infertility. If detected during childhood, therapy should commence early in the second decade to simulate normal puberty and avoid the adult manifestations of the disorder. Cases with XY/XXY mosaicism should be individually evaluated regarding the advisability of testosterone therapy.

XXXXY syndrome. True to the general rule the presence of the Y chromosome confers maleness on such individuals. However, the presence of three extra X chromosomes appears responsible for many somatic and genital anomalies [17]. Intellectual and emotional retardation is also prominent. Short stature, sternal anomalies such as pectus excavatum and limb malformations, e.g. radio-ulnar synostosis, hypertelorism, mongoloid slant of the

---

Fig. 3. a, b Buccal smear and sex chromosome complement of Klinefelter syndrome showing X chromatin positive nucleus and two X chromosomes. *c, d* Sex chromosome complement (with autoradiography) and buccal smear on tetra XY male – note three late replicating X chromosomes and three X chromatin bodies. *e, f* Karyotype in Down's syndrome in a male and female patient, respectively, the latter with G-banding. *g* Deletion of the short arm of chromosome No. 5 in cri-du-chat syndrome.

Genetic Aspects of Male Infertility

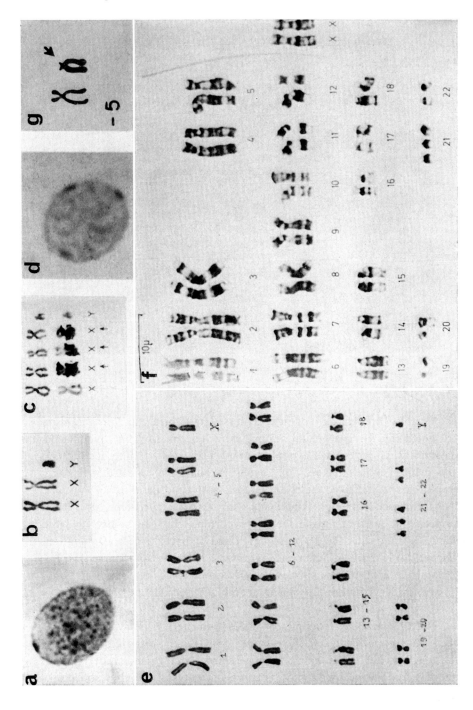

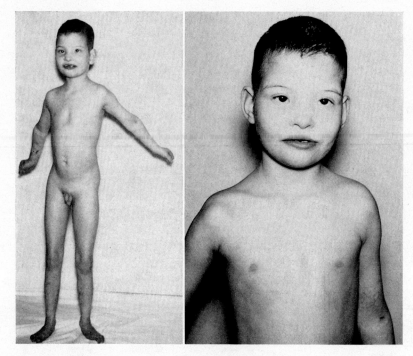

Fig. 4. An 8-year-old boy with XXXXY (tetra XY) syndrome.

palpebral fissures, epicanthic folds, depressed nasal bridge and prognathism are noted (fig. 4). The genital anomalies include micro-penis, hypoplastic scrotum, small testes with hypoplastic and hyalinized seminiferous tubules and reduced interstitial cells of Leydig. A quarter of affected males are cryptorchid. All affected males are infertile with azoospermia.

Diagnosis: Although individuals with XXXXY syndrome have been suspected at birth of having Down's syndrome, this can be excluded on clinical grounds and the diagnosis confirmed by buccal smear examination where three X-chromatin masses in interphase nuclei are discernible. Karyotyping will clinch the diagnosis when 49 chromosomes are discovered and three extra X chromosomes can be unambiguously identified (fig. 3c, d).

XXXY syndrome. Examples of males with the XXXY syndrome have been encountered in many centers [49]. They resemble the more familiar XXY Klinefelter syndrome but are of shorter stature and more marked retardation. The behavior may be more deviant and possibly antisocial.

They are all hypogonadic and infertile. The testes are small and tend to be fibrotic in consistency. On biopsy, hyalinized ghost tubules are encountered, lacking in germinal epithelium and lined only by Sertoli cells.

Diagnosis: Clinical suspicion should be confirmed by buccal smear examination where two X-chromatin bodies will be seen. Chromosome studies will reveal a karyotype of 48, with XXXY sex chromosome make-up.

Etiology: Non-disjunction affecting both paternal and maternal gametogenesis is invoked as a possible explanation. Non-disjunction affecting the first meiotic division during spermatogenesis and either meiotic division during oogenesis will result in an XY bearing sperm and XX bearing ovum respectively. Fertilization of the latter by the former would result in a zygote with a total chromosome count of 48 and XXXY sex chromosomal complement. Successive non-disjunction in either parent alone may lead to the same ultimate outcome. A normal Y-bearing sperm may fertilize an XXX-bearing egg or an XXY-bearing sperm may fertilize a normal X-bearing egg with the resulting zygote having a total of 48 chromosomes with XXXY sex chromosomal complement.

Management of XXXXY and XXXY syndrome: Infertility is incurable. Androgen replacement therapy is recommended after due consideration is given to the emotional and intellectual make-up of the affected individual.

Numerical Anomalies of the Y Chromosome

The four main numerical abnormalities of the Y chromosome are the presence of an extra Y chromosome as in the XYY syndrome; the presence of two Y chromosomes in addition to two X chromosomes in the XXYY syndrome; the absence of the Y chromosome in a cell line of male mosaic individual where the chromosomal make up is XY/X0 in the two cell lines; and the absence of the Y chromosome in the XX male.

The XYY syndrome. The disorder affects 1 in 750 newborn males. The features which usually become evident in childhood include tall stature with increased length versus breadth of bones [58], sternal anomalies such as pectus excavatum, large teeth and bossing of the skull [19]. The mentality is dull and impulsive behavior with low frustration tolerance and aggressive outbursts are characteristic [53]. The latter may be directed against person or property with the former particularly having sexual overtones [14]. The probability of a serious offence is significant and likelihood of landing in an institution for male juvenile delinquents or later in a maximum security penitentiary has been estimated at 30-fold that of XY males [37].

Cryptorchidism, small penis and hypospadias have been observed relatively more often among XYY than XY individuals [18]. The fertility of the XYY males is therefore diminished at least in part due to the concomitant genital anomalies.

Diagnosis: May be clinically suggested in infancy but is more often overlooked during childhood and even adulthood. If suspected, a buccal smear examination will reveal two highly fluorescent Y-chromatin bodies (with UV microscopy) in the nuclei. These correspond to the heterochromatic portions, particularly the long arms of the Y chromosomes. This investigation of buccal smears may be done along with the more usual examination, using ordinary stains and light microscopy, for the X-chromatin (Barr) bodies. The stain used is a quinacrine derivative which causes fluorescence of the Y chromosome and segments of other chromosomes under UV light. Eventual confirmation will, of course, come from karyotypic analysis where a total of 47 chromosomes will be found with two acrocentric highly fluorescent Y chromosomes.

XXYY males. The patients tend to have features similar to XXY Klinefelter syndrome in terms of tall stature. They usually exceed in height individuals with either the XXY Klinefelter and XYY syndrome. They are hypogonadic and infertile with small testes with hyalinized tubules and Leydig cell hyperplasia. The intelligence is low and the mood is temperamental with aggressive tendencies [37]. The habitus is eunuchoid and gynecomastia is prevalent.

Diagnosis: Clinical suspicion should be followed by buccal smear examination where two fluorescent Y bodies and one X-chromatin body will be discerned. Karyotype will confirm the presence of 48 chromosomes with XXYY sex chromosomal complement.

Treatment: Testosterone administration is probably inadvisable in view of the consistent antisocial behavior and low intellect of affected individuals. Non-virilizing preparations to improve nitrogen retention may be a more suitable substitute.

XY/X0 mosaicism. Cryptorchidism, hypospadias, micro-penis and persistent urogenital sinus have all been observed. The degree of involvement corresponds to the proportion of the X0 to the XY cell lines, and its distribu-

---

Fig. 5. Ambiguous genitalia and karyotype in X0/XY syndrome.

Genetic Aspects of Male Infertility

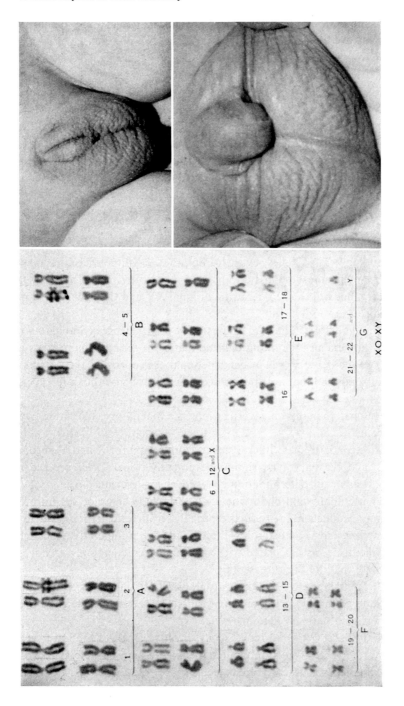

tion in the various body sites, in particular the gonads and genital tissue. Most of these males are hypogonadic and all are infertile.

Among the various entities of ambiguous genitalia, this is the one in which determination of sex and assignment of gender may prove hardest.

Diagnosis: The clinical suspicion is often aroused by ambiguous or abnormal genitalia. The diagnosis is suggested by the finding of a low frequency of fluorescent Y-chromatin body in the buccal smear and confirmed by karyotype. Two lines will be detected, one with 45 chromosomes and a single X chromosome (Turner-like) and another with a normal male karyotype (46, XY) (fig. 5). The ratio of the two lines should be determined using a minimum of 100 nuclei. Furthermore, in these instances more than a single tissue should be sampled, for example, blood and skin (from more than one location, preferably involving the genital area). If a biopsy or excision of the gonads is carried out, they should be karyotyped as well.

Etiology: The likeliest event leading to XY/X0 mosaicism is anaphase lag with Y chromosome loss during early mitotic division of a male zygote. Two cell lines will thus co-exist, the original XY and the new X0 cell populations.

Ambiguous genitalia management: The management of ambiguous genitalia at birth is an urgent and delicate matter. If serious doubt exists as to the sex of the baby, no sex or gender should be reported to the parents who should be advised that a definitive decision will soon (within a few days) be reached.

The first four sets of investigations are mandatory, the latter two are elective:

(1) Genetic studies: Buccal smear examination for X and Y-chromatin bodies, and the karyotype should be urgently done. These are the most crucial tests in sex determination and gender assignment.

(2) Metabolic and endocrine studies: Include assay of sex hormones, their intermediate and byproducts especially serum testosterone, 17-hydroxy-progesterone and 17 ketosteroids. Electrolyte assay is imperative since certain metabolic disorders associated with ambiguous genitalia, e.g. adrenogenital syndrome, may be life threatening.

(3) Radiography: Excretory urography (IVP) and urethrocytography to exclude concomitant urinary malformations and define the structures of the lower genitourinary system.

(4) Endoscopy: Urethrocytoscopy directly visualizes the lower genitourinary tract and dectets anomalies therein. It may be deferred for a few weeks unless more urgently indicated.

Genetic Aspects of Male Infertility

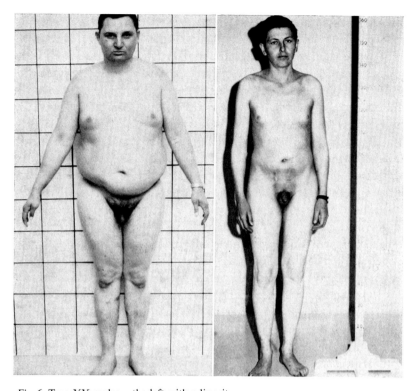

Fig. 6. Two XX males – the left with adiposity.

(5) Gonadal biopsy: The information derived from the above is usually sufficient to determine the sex and assign the appropriate gender. In few cases, biopsy of an accessible gonad may be justified. Even then some cases may defy proper assignment when a mixture of ovarian and testicular tissue is encountered (ovitestis).

(6) Laparoscopy and laparotomy: Either laparoscopic or surgical exploration of the abdomen for direct inspection of the internal genitalia and gonads and biopsy of the latter may be rarely done in cases of ambiguous genitalia to settle the sex of the individual. Laparoscopy is impractical in the newborn leaving laparotomy as the only alternative.

The XX male. Over 60 cases are now on record of males with the XX sex chromosomal complement [22]. This conflicts with the rule of the Y chromosome determining maleness and its absence determining femaleness. The phenotype among those males is variable with the majority resembling males with the XXY Klinefelter syndrome (fig. 6, left) except for their relatively short stature (fig. 6, right) and normal intelligence. They are somewhat hypogonadic and are all infertile. Testicular biopsy done on many revealed no spermatogenesis with hypoplastic hyalinized and fibrosed tubules, and pleomorphic interstitial cells of Leydig [23]. Few patients were born with ambiguous genitalia whereas rare examples may even be classified as hermaphrodites with mixed gonadal dysgenesis.

Etiology: Three mechanisms may account for XX human males: (a) nonhomologous translocation involving the X and Y chromosomes. The male determining genes are believed to be located on the short arm of the Y chromosome. Through translocation during the first meiotic division in the father these may come to reside on the X chromosome [28]; (b) undetected mosaicism whereby the Y chromosome in a stem line guided embryonic testicular differentiation but since disappeared or remained closely circumscribed; (c) the gene theory. Male differentiation is triggered by a locus on the short arm of the Y chromosome. The locus is either the same or very closely linked to that determining the H-Y histocompatibility antigen [84]. All XX males thus examined were H-Y positive [83]. This suggests that either the responsible locus was translocated to the X chromosome or an autosome, or mutation in an autosomal locus directed the differentiation towards the male sex despite the absence of the Y chromosome. The mutation, which corresponds to similar disorder in goats [32] and mice [15], has been termed sex reversal gene (vide infra). Its inheritance may be autosomal dominant with limitation to XX individuals, though recessive transmission cannot yet be ruled out [23].

Management: Infertility which is uniform in XX males is incurable. Hypogonadism could be improved by testosterone replacement therapy.

Structural Anomalies of the Y Chromosome

Two major lesions may affect the Y chromosomes with significant impact on the sexual development and fertility of the individual.

Deletion. Early loss of a fragment of the Y chromosome may lead to failure or normal male sexual differentiation. Studies on individuals with deletion of the short arm of the Y chromosome suggest that the masculinization genes largely reside there [27, 38]. The long arm may also be implicated in testicular differentiation [71].

Di-centric Y chromosome. As a result of complicated cytogenetic events, two Y chromosomes in an XYY zygote may undergo deletion of portions of

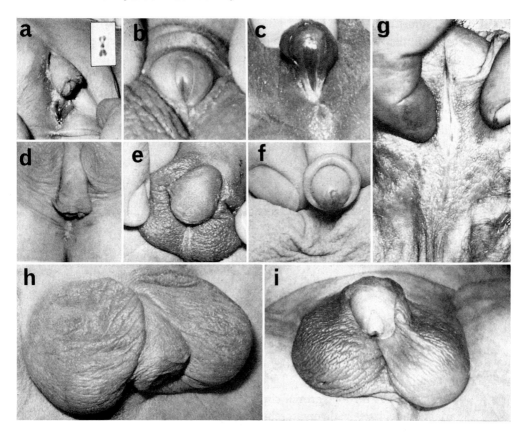

Fig. 7. External genitalia. *a* X dicentric Y syndrome (note dicentric Yp chromosome insert). *b* Deletion long arm of chromosome No. 18 (18q-). *c* Pseudovaginal perineoscrotal hypospadias syndrome. *d* 3β-hydroxysteroid dehydrogenase deficiency in a male. *e* 21 steroid hydroxylase deficiency in a female (female pseudohermaphroditism). *f* Hypogonadotrophic eunuchoidism. *g* Telecanthus-hypospadias syndrome. *h* Incomplete pseudohermaphroditism in a male. *i* Reifenstein syndrome.

either their long or short arms. The two acentric fragments are lost, the remaining centric portions of the Y chromosomes join to form a di-centric Y chromosome (fig. 7a). Such individuals often display both hypogonadism and infertility and many are born with ambiguous genitalia (fig. 7a) [4].

Disorders of the Autosomes

Although not directly involved in the determination of sex, abnormalities of the autosomes may lead to infertility.

Numerical Abnormalities

Disorders which do not permit survival into the reproductive period, such as trisomy D or trisomy E, are obviously incompatible with fertility.

Down's syndrome. Whereas female patients may be fertile males with trisomy 21 (fig. 3e, f) are characterized by absolute infertility with azoospermia in the male. No spermatogenesis takes place in their testes and hyalinization and fibrosis of the seminiferous tubules is the rule [65]. Furthermore, they display evidence of substantial hypogonadism.

Trisomy 8. Most of the patients identified to date have been mosaics. Cryptorchidism with consequent azoospermia is the usual finding in males [70].

Structural Anomalies

Structural anomalies of the autosomes are frequently associated with male infertility as a part of a constellation of multiple physical and mental abnormalities.

Cri-du-chat syndrome. Deletion of the short arm of chromosome No. 5 (fig. 3g) is characterized by severe mental retardation and multiple physical abnormalities including bilateral cryptorchidism, azoospermia and infertility in the male [10].

Deletion of the long arm of chromosome No. 18 (fig. 7b) [21] has been variably associated with cryptorchidism, micro-penis, impotence and sterility due to azoospermia.

Single Gene Disorders

When the disorder is due to a genetic error confined to a single locus (or a pair of homologous loci) typical Mendelian inheritance prevails. The transmission of the trait depends on whether the relevant locus is on an autosome or the X chromosome. In general, in the former, it will manifest equally in both sexes, whereas if X-linked, difference in expression in the two sexes will be apparent. The second major criterion is whether a pair of mutant alleles is necessary for the manifestation of the disorder or a single mutant allele is sufficient. In the former case the trait is called recessive, in the latter dominant. Genetic disorders are thus described as autosomal dominant or recessive and X-linked dominant or recessive. Several conditions representing various modes of inheritance are associated with infertility in the male.

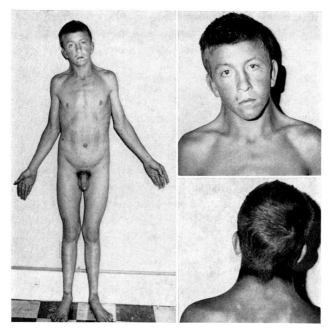

Fig. 8. Noonan syndrome.

Autosomal Dominant Disorders

Noonan Syndrome (fig. 8)

The disorder is characterized by mild to moderate mental retardation, short stature, epicanthic folds, bilateral ptosis, strabismus and hypertelorism, prominent low set auricles, low nuchal hair line and neck webbing. Shield-like chest, widely spaced nipples and pectus excavatum, hemi-vertebrae and incomplete neural arches, cubitus valgus and genu valgum are also noted. Often there is a congenital heart lesion, either pulmonic stenosis or septal defect [59]. The genitalia are small with frequent cryptorchidism and hypospadias. The condition which affects both sexes equally is often associated in the male with hypogonadism and infertility.

Etiology: The disorder appears to be inherited as a single gene defect in an autosomal dominant fashion [8, 67].

Myotonic Dystrophy (Steinert Disease)

Features include baldness, cerebral atrophy, intellectual and emotional impairment, cataract and conductive cardiac lesions in addition to the muscle involvement characterized by myotonia, muscle weakness and wasting [16]. Gonadal dystrophy afflicts both sexes, but more so the male. Both the spermatogenic and endocrine elements in the testes are involved ultimately resulting in infertility and impotence.

Sex Reversal Syndrome

Familial examples of males with XX sex chromosomal make-up have been reported [41]. A male determining autosomal dominant gene may be transmitted from father to offspring expressing itself only when the sex chromosomal make-up was XX, i.e. sex limitation to chromosomal females, akin to the dominantly transmitted sex reversal syndrome in mice [15]. The autosomal gene appears to guide the differentiation of the indifferent gonad (in the chromosomal females) towards testicular development. The disorder represents a genetically determined variant of XX males [22] (vide supra).

Achondroplasia

This major cause of dwarfism is recognized as an autosomal dominant disorder. Only one eighth of all cases seen are, however, found to be transmitted from either parent, the remainder of the cases being the result of fresh dominant genetic mutations. Advanced paternal age increases the probability of such mutations [64]. This reflects reduced fertility (reduced genetic fitness) of patients with achondroplasia, between 10 and 15% that of unaffected males [55].

Autosomal Recessive Disorders

Cystic Fibrosis

The most common simply inherited serious disorder, occurring in approximately 1 of 2,000 newborns, is associated with nearly complete sterility in the male. Failure of normal development of the vas deferens was suggested as the cause of infertility [40]. More likely obstruction of the transport ducts of male genital system, not unlike that in the pancreas and salivary glands, is responsible [63]. Paradoxically heterozygotes are believed to have higher fertility than controls [20].

Unlike its genetics, the molecular basis for cystic fibrosis has not been elucidated. Possible areas of involvement include enzyme deficiency such as arginine esterase [68], structural or functional alteration or absence of a serum protein component such as α_2-macroglobulin [74], and the presence of abnormal moieties such as the ciliary inhibitory factor [78].

Congenital Anorchia

Absence of the testes either bilateral or unilateral has been reported in brothers [7] and identical twins [31]. The males have normal chromosomal make-up with 46 total number and XY sex chromosome complement. These unmistakable males always acquire the male gender. Though mistakenly diagnosed as cryptorchidism, the vas deferens ends blindly. The findings suggest that a catastrophe such as torsion or vascular occlusion befell the testes during their descent, hence the term 'vanishing testes'. The apparently normal male phenotype is evidence for a late fetal event permitting the testicular function to achieve normal virilization [1].

The familial distribution and concordance in identical twins favor a genetic etiology while not precluding a fetal accident as the pathogenesis. Autosomal recessive transmission with sex limitation is likely although X-linkage has not been ruled out.

Bilaterally affected males behave like castrates and their management consists of androgen administration to enhance masculinization. They are, however, incurably infertile.

Hypertelorism, Cryptorchidism and Skeletal Anomalies

In this autosomal recessively inherited disorder hypertelorism, cryptorchidism and infertility are associated with skeletal anomalies. These include digital contractures, sternal deformity with premature fusion of the manubrium and corpus sterni and multiple osteochondritis dissecans involving the knees, elbows, calcanei and spine [33].

Pseudovaginal Perineoscrotal Hypospadias (fig. 7c)

The disorder manifests with ambiguous genitalia: small phallus, a blind-ending perineal opening resembling a vagina and severe hypospadias with the urethral opening at the junction of the penis and a bifid scrotum [61]. At puberty such individuals exhibit masculinization including male habitus, excellent muscular development, phallus enlargement, voice change and semen production, but scant beard growth. The prostate remains small and they are infertile. They do not display gynecomastia. Although often brought up as girls, by the time of puberty their male sex becomes evident. Psychosocial conflict between their newly determined sex and gender (based on rearing) may, therefore, ensue. The internal genitalia are male with no significant Mullerian but fully differentiated Wolffian system. Moreover, the testes are ostensibly histologically unremarkable.

The conversion of testosterone which is of normal level to 5α-dihydrotestosterone (5αDHT) is defective, probably due to deficient 5α-testosterone reductase [85].

The responsible locus likely controls the synthesis of the enzyme. The evidence for sex-limited autosomal recessive transmission includes the presence of more than one sibling with the condition, freedom of the parents of similar manifestations, approximately 1:8 recurrence risk (1:4 for autosomal recessive inheritance × 1:2 for the male sex), relatively high consanguinity rate among the parents of patients [24], strong tendency to clustering among certain ethnic groups and geographically isolated populations, e.g. the Dominican Republic [36] and normal male karyotype.

Persistent Mullerian Duct Syndrome

Affected males show bilateral cryptorchidism and inguinal herniae. Internally a uterus and Fallopian tubes are found. Their gonads are testes with arrested spermatogenesis but Leydig cell function. The abnormality is attributed to failure of the manufacture, or action of an early embryonic testicular factor recently identified as a polypeptide which causes regression of Mullerian duct development, hence the persistence of their derivatives – the uterus and Fallopian tubes. Consanguinity among the unaffected parents of affected brothers suggest autosomal recessive inheritance [5]. The disorder is, of course, limited to the XY male individuals with a recurrence risk of 1:8 among the patient's siblings.

Congenital Adrenal Hyperplasia (Adrenogenital Syndrome) [9]

These autosomal recessive disorders have been numbered I through V to signify the order of the steps in the synthetic pathway. Table I summarizes the relevant features.

Familial Hypogonadotrophic Eunuchoidism

The disorder which affects both sexes appears to be transmitted in an autosomal recessive fashion with a pair of carrier parents who are more often related than by chance alone. One or more siblings may be affected [45].

Paucity of secondary sexual characteristics and eunuchoid build with long extremities relative to the trunk, cryptorchidism and small penis are due to an isolated deficiency of pituitary gonadotrophin (fig. 7f) [79]. Unlike clomiphene which is ineffective [26], testosterone treatment brings about a dramatic response [35]. Infertility is, however, invariable and incurable.

Table I. Congenital adrenal hyperplasia

Type	Deficient enzyme	Metabolic block	Clinical features	Main steroid secreted	Urinary 17-keto-steroids	Prognosis
I	20,21-desmolase	cholesterol to pregnenolone failure of adrenocortical and testicular hormone synthesis	hypogonadism, male pseudohermaphroditism, feminine external genitalia, severe salt and water wasting	cholesterol?	low	often fatal
II	3β-hydroxysteroid dehydrogenase (fig. 7d)	pregnenolone to progesterone block of adrenocortical and testicular hormone synthesis	hypogonadism, male pseudohermaphroditism, feminine external genitalia, severe salt and water wasting	pregnenolone and 17-hydroxy-pregnenolone (3β-hydroxy-steroids)	elevated	often fatal
III	17-hydroxylase	diminished cortisol, testosterone, aldosterone synthesis; excessive estrogen and corticosterone and deoxycorticosterone production	hypogonadism and infertility, male pseudohermaphroditism, ambiguous genitalia [57], gynecomastia, hypertension, hypokalemic alkalosis and hypervalemia, no salt loss	deoxycorticosterone and corticosterone	low	treatable, compatible with survival and reproduction
IV	21-hydroxylase	diminished glucocorticoid and mineralocorticoid synthesis, excessive adrenocortical androgen production	virilization, precocious male puberty (pseudohermaphroditism) (fig. 7e), in females, male infertility is common; salt and water loss in approximately 1/3 of patients	17-hydroxy-progesterone	elevated	treatable, compatible with survival and reproduction
V	11-hydroxylase	diminished cortisol and aldosterone synthesis, excessive adrenocortical production	virilization, precocious male puberty pseudohermaphroditism in females; male infertility is common; hypertension	11-deoxy-cortisol (compound S), desoxycorticosterone	elevated	treatable, compatible with survival and reproduction

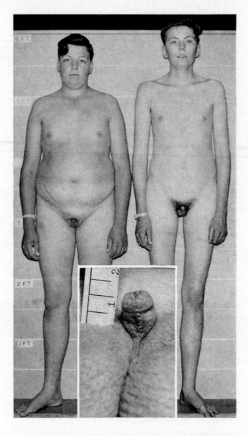

Fig. 9. Two brothers with familial hypogonadism and infertility – note small penis.

Hypogonadism with Microcephaly

Hypogonadism together with syndactyly, microcephaly, and low grade mental retardation has been observed among three brothers whose parents were consanguineous. No spermatogenesis was demonstrable on testicular biopsy and the interstitial cells were replaced by fibrous tissue. The affected males were, of course, infertile.

Hypogonadism and Infertility

Many male patients are seen whose only demonstrable pathology is hypogonadism and its sequelae and infertility. Frequently more than one brother is affected (fig. 9) and the parents are related – favoring autosomal recessive inheritance. X-linked recessive inheritance is, however, a feasible alternative. The infertility is incurable but hypogonadism can be alleviated by adrogen andministration.

X-Linked Dominant Disorders

Telecanthus-hypertelorism hypospadias syndrome appears to be transmitted in an X-linked dominant fashion with males being more severely affected. Apart from the evident facial appearance, hypertelorism, maxillary recession, prognathism, cleft palate, bifid uvula and mental retardation, hypospadias of various degrees of severity (fig. 7g), cryptorchidism, and other perineal anomalies such as anterior displacement of the anus occur in this disorder. More complicated urinary malformations have also been noted [69]. The infertility, though inconsistent, is common. Oligospermia is practically invariable and azoospermia is frequent.

X-Linked Recessive Disorders

Testicular Feminization Syndrome

The disorder is confined to chromosomal males (46 with XY sex chromosomal complement). Two alternate modes of inheritance exist: X-linked recessive and autosomal dominant with limitation to the male (XY) sex. The former is more likely. Evidence from affected families [29], animal model in the mouse [48, 60] and rat [6, 11] make it almost certain that X-linked recessive transmission operates.

The underlying mechanism has been shown to be the absence of both testosterone and dihydrotestosterone receptors in the cytosol. This determines end-organ unresponsiveness with absence of masculinization despite normal or even somewhat increased testosterone production. For want of masculinization, the external genitalia remain feminine. The internal genitalia (derivatives of the Wolffian ducts) are male-like in structure. The estrogen output by the Sertoli cells of the testes is increased with secondary sexual characteristics reminiscent of normal females (fig. 10). The feminine configuration, and ample breast development led some of them to be models [56], stewardesses [51], strippers or even prostitutes [66]. They lack both axillary and pubic hair – features dependent in both sexes on response to testosterone.

The molecular basis and X-linked recessive inheritance have been con-

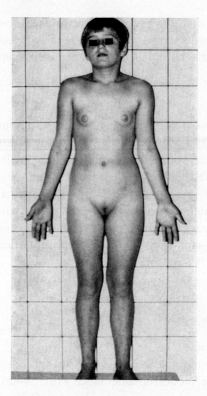

Fig. 10. A chromosomal male (XY) with complete testicular feminization.

firmed by finding two clones of fibroblasts in obligate female heterozygotes, one with androgen-binding receptors and one without [52]. Genetic heterogeneity in testicular feminization has been demonstrated [3]. All affected individuals are infertile.

Incomplete Testicular Feminization

Shades of testosterone responsiveness exist between normal male response to complete refractoriness as seen in testicular feminization. In its incomplete form, patients manifest external features characteristic of both sexes; for example, breast development, clitoral enlargement, labial fusion, some virilization at puberty, and some pubic and axillary hair growth. Some spermatogenesis [47], partial androgen binding [42], and response [86] have been noted. Most such affected individuals are raised as females and none is fertile. Inheritance is X-linked recessive.

Complete Male Pseudohermaphroditism

The features include perineoscrotal hypospadias, absent vas deferens, undescended testes, and vaginal orifice. It occurs in chromosomal males (XY) and is inherited as an X-linked recessive trait.

Incomplete Male Pseudohermaphroditism (fig. 7h)

This appears to be a milder version of the above disorder with hypospadias, small phallus and bifid scrotum, gynecomastia and generalized evidence of hypogonadism occurring [81]. The karyotype is 46, with XY sex chromosome complement.

Reifenstein Syndrome (fig. 7i)

It occupies a position intermediate in severity between the previous two disorders. The features of this form of male pseudohermaphroditism are hypospadias, hypogonadism, gynecomastia, normal male karyotype (46, XY), and frequent familial distribution compatible with X-linked recessive transmission. On testicular biopsy germ cells with mitotic and even meiotic activity could be demonstrated. No spermatozoa are, however, encountered and the affected males are infertile. Leydig cell hyperplasia and hyalinized seminiferous tubular ghosts attest to excessive gonadotrophic activity.

Incomplete male pseudohermaphroditism, Reifenstein syndrome, complete male pseudohermaphroditism, partial and complete testicular feminization probably represent progressively decreasing androgen responsiveness with the complete testicular feminization displaying an apparent female phenotype. The disorders which are inherited in an X-linked recessive fashion may all be determined by various alleles at the same X-linked locus [30].

Kallmann Syndrome
(Hypogonadotrophic Hypogonadism with Anosmia) [39]

Affected males show hypogonadism due to low gonadotrophic production. Anosmia due to agenesis of olfactory lobes bulbs and tracts is the other major feature. The defect involves the functions of both the pituitary and interstitial cells of Leydig with deficiency in the secretion of both FSH and LH gonadotrophic hormones as well as insensitivity of the Leydig cell response to the gonadotrophins. Failure of response to clomiphene as ascertained by plasma levels of gonadotrophins was claimed [73]. The affected males have small genitalia, eunuchoid build, diminished sexual hair and gynecomastia. On testicular biopsy few germ cells with arrest of spermatogenesis at the primary spermatocyte stage are discerned. Leydig cells are very sparse or absent. The syndrome has been ascribed to a disorder of hypothalamic regulation [50]. Although infertility is the rule, there are instances of affected males who reproduced [34].

X-linked recessive transmission is the likely mode of inheritance. Heterozygous females manifest partial expression with anosmia or diminished secondary sexual characteristics such as delayed menarche, irregular menstruation, and poor breast development. More severe expression includes primary amenorrhea and persistent fetal ovaries [77]. The variability of expression is in favor of X-linked inheritance and can be readily explained on the basis of random X chromosome inactivation (lyonization).

Del Castillo Syndrome

Germinal cell aplasia (Sertoli-cell-only syndrome): The disorder is characterized by obesity, gynecomastia, and oligospermia at adolescence progressing to azoospermia by adulthood when gynecomastia often disappears [25].

Gonadal Dysgenesis XY Female Type

In these chromosomal males, only streak gonads are noted. The internal genitalia are female but, of course, they do not ovulate nor spontaneously menstruate. The patients have normal stature and lack the stigmata of X0 Turner syndrome [72]. Neoplasia, which includes dysgerminomas and gonadoblastomas may affect the streak gonads [82]. Several males who were maternally related have been described [80]. Although X-linked recessive inheritance is the most likely mechanism, autosomal dominance with limitations to XY individuals is a plausible alternative.

Polygenic or Multifactorial Disorders

When no single identifiable major locus is found to determine a genetic disorder, its etiology is then the result of the collective action of several gene loci. It is described as multifactorial with polygenic genetic component – the remainder, if any, being the contribution of the environment.

Many genetic disorders which cause partial or complete infertility in the male are inherited in this fashion. Although they do not follow the simple Mendelian rules of inheritance. The risks of recurrence among first and second degree relatives, brothers and sons, and grandsons, nephews and uncles, respectively, are higher than the random risk in the general population. This familial clustering, higher concordance in monozygotic than dizygotic twins, disparity in prevalence among racial and ethnic groups sharing the same environment are evidence for polygenic inheritance. Some disorders are generalized whereas others are localized in their manifestations.

Generalized Polygenically Determined Disorders

Diabetes mellitus

Although controversy still exists as to the mode of inheritance of juvenile diabetes, the present evidence strongly suggests that maturity-onset diabetes mellitus is inherited as a multifactorially or polygenically determined trait. Relative infertility affects male diabetics, with impotence as a possible cause in some patients. Since diabetics frequently have the onset of the clinical disease relatively later in life, the influence of partial infertility may be minor.

Thyrotoxicosis and Hypothyroidism

Both disorders may impair fertility in both sexes and potency in the male. Although simple Mendelian inheritance operates in some, the majority are polygenic.

Localised Disorders

Cryptorchidism

When cryptorchidism is an isolated feature, the genetic etiology is likely polygenic. When it is a feature of a multifaceted syndrome, Mendelian inheritance might operate as in Noonan's syndrome which is autosomal dominant, or pseudovaginal perineoscrotal hypospadias which is autosomal recessive. Alternatively, it may be a feature of chromosomal syndrome as sometimes seen in Down's syndrome and always in cri-du-chat syndrome. Cryptorchidism may be a feature of a syndrome of unknown etiology such as the triad (abdominal muscle absence, urinary tract anomalies and cryptorchidism) or prune belly syndrome. It may also represent a fresh autosomal dominant mutation. Uncorrected (and in many instances corrected) cryptorchidism is incompatible with male infertility.

Hypospadias

Significant hypospadias constitutes a component of a complex disorder such as the autosomal dominant Robinow syndrome, the autosomal recessively inherited Smith-Lemli-Opitz syndrome [76] or the X-linked dominantly inherited hypertelorism-hypospadias syndrome [69]. Occasionally, isolated hypospadias of a mild degree may follow autosomal dominant transmission [75]. More frequently it is determined in a multifactorial fashion with polygenic genetic etiology. It may cause infertility which is surgically reversible.

Meningomyelocele (Spina bifida)

Both spina bifida manifesta and occulta are attended with frequent infertility in the male. Although spermatogenesis may occur in many instances, the transport, storage and ejaculation of these sperms may be impaired. Often retrograde ejaculation into the bladder can account for the demonstrable azoospermia. The disorder is multifactorial in etiology with a polygenic genetic component. It belongs to the group of disorders of defective closure of the neural tube. These encompass, besides meningomyelocele with or without hydrocephalus, meningocele, anencephaly and meningoencephalocele.

Fertility and Genes

So far, we have considered the influence of chromosomes and their genes on fertility. It is apt, however, to consider the impact of fertility on the genes and their frequency. The frequency of a particular form of a gene (allele) in the population is the proportion of that allele to all the possible alleles at the locus. The frequency of an allele in future generations depends largely on its frequency in the present generation and the fitness of those individuals who carry it.

Fitness in this context is a strictly relative affair which is rather narrowly defined. It does not imply physical strength, endurance, wealth, power, influence, wisdom or courage, although all these may indirectly contribute to it. Genetically defined, fitness is the relative contribution of the individual in terms of the number of viable offspring. It is a ratio with a denominator of 1 (unity) as the average fitness in the population. Specific genes may confer more or less fitness than 1 on their carriers. If the fitness is more than 1 the gene has a selective advantage. Barring a loss of this advantage due to a major environmental alteration, it will progressively increase in prevalence until it approaches 100%. At that point it will have virtually eliminated all other alternate genes (alleles) and will be said to have been 'fixated'. Conversely, a gene which confers below average fitness on its carriers has a selective disadvantage. It will gradually decline in frequency until it disappears – or is 'eliminated'. This inexorable trend will continue unless the disadvantage is alleviated or reversed either due to natural or artificial environmental change, e.g. therapeutics. Genes influence fitness through differential mortality or survival and differential fertility. An individual who is dead cannot, of course, reproduce. His genes will not be transmitted to future generations, if his death precedes his reproductive age. Homozygotes for thalassemia or sickle cell anemia and male hemizygotes for Duchenne muscular dystrophy may fall in this category. Their genetic fitness is zero. They have, therefore, been referred to as examples of genetic death. With their death their genes will disappear from the population gene pool without being carried by successors. The gene frequency may fall to a critical level at which by random fluctuation or chance alone it may disappear. Another allele at the same locus will conversely rise in frequency due to the differential survival of its bearers.

More pertinent to this chapter is the role of fertility in influencing gene frequency. A male who carries the gene for testicular feminization has essentially normal survival but is invariably sterile. The gene will tend to decline in frequency until it reaches an equilibrium at a very low level in which recurrent mutations will counterbalance the loss due to infertility.

In less extreme examples of infertility in which the genetic fitness is less than 1 but more than zero, the affected individuals will contribute less offspring than average to the future generations. An equilibrium is established by the opposing forces of mutations and negative selection in which the responsible gene(s) may not disappear but will be maintained at a low frequency.

In contrast, individuals who have an allele that confers on them higher than average survival or fertility will have genetic fitness above 1. They will tend to outreproduce, on average, the remainder of the population. Examples include heterozygotes for cystic fibrosis or for sickle cell anemia (sickle cell trait). In the latter individuals conclusive evidence has been submitted indicating greater resistance over normal homozygotes, towards infection with falciparum malaria – a major killer in endemic areas [2].

The situation has been named balanced polymorphism. It is believed to be an important genetic mechanism for perpetuation of genes which are lethal or detrimental in the homozygous state. The selective advantage (through higher than average genetic fitness) of heterozygotes counterbalances the disadvantage to affected homozygotes. It depends on a particular genetic environment interaction such as that of sickle cell hemoglobin and falciparum malaria, whose eradication will abolish this advantage with quick decline in gene frequency. The specific environmental interaction which favors the fitness of heterozygotes for cystic fibrosis has hitherto eluded elucidation.

Conclusion

The above review illustrates the broad spectrum of abnormalities which are, either in whole or in part, genetically determined and are associated with infertility in the male. The latter is often absolute but in a few disorders may be only relative. Although in a minority of conditions, e.g. adrenogenital syndrome, it may be averted, the majority of infertility of genetic origin is irreversible. When hypogonadism is simultaneously encountered, it, however, can often benefit from androgen replacement therapy. Elucidation of the mechanism of infertility in some of these disorders may shed light on possible avenues of forestalling or treating it.

References

1 ABEYARANTNE, M.R.; AHERNE, W.A., and SCOTT, J.E.S.: The vanishing testis. Lancet *ii:* 822–824 (1969).
2 ALLISON, A.C.: Polymorphism and natural selection in human populations. Cold Spring Harb. Symp. quant. Biol. *29:* 137–149 (1964).
3 AMRHEIN, J.A.; MEYER, J.W.; JONES, H.W., jr., and MIGEON, C.J.: Androgen insensitivity in man: evidence for heterogeneity. Proc. natn. Acad. Sci. USA *73:* 891–894 (1976).
4 ARMENDARES, S.; BUENTELLO, L.; SALAMANCA, F., and GANTU-GARZA, J.M.: A dicentric chromosome without evidence of sex chromosomal mosaicism. 46, XYq dic in a patient with features of Turner's syndrome. J. med. Genet. *9:* 96–100 (1972).
5 ARMENDARES, S.; BUENTELLO, L., and FRENK, S.: Two male sibs with uterus and Fallopian tubes. A rare probably inherited disorder. Clin. Genet. *4:* 291–296 (1973).
6 BARDIN, C.W.; BULLOCK, L.; SCHNEIDER, G.; ALLISON, J.E., and STANLEY, A.J.: Pseudohermaphrodite rat: end organ insensitivity to testosterone. Science *167:* 1136–1137 (1970).
7 BOBROW, M. and GOUGH, M.: Bilateral absence of testes. Lancet *i:* 366 (1970).
8 BOLTON, M.R.; PUGH, D.M.; MATTIOLI, L.F.; DUNN, M.I., and SCHIMKE, N.: The Noonan syndrome: a family study. Ann. intern. Med. *80:* 626–629 (1974).
9 BONGIOVANNI, A.M.: Disorders of adrenocortical steroid biogenesis; in STANDBURY, WYNGAARDEN and FREDRICKSON The metabolic basis of inherited disease; 3rd ed., pp. 857–885 (McGraw-Hill, New York 1972).
10 BREG, W.R., *et al.:* The cri-du-chat syndrome in adolescents and adults. J. Pediat. *77:* 782 (1970).
11 BULLOCK, L.P. and BARDIN, C.W.: Androgen receptors in testicular feminization. J. clin. Endocr. Metab. *35:* 935–937 (1972).
12 CALDWELL, P.D. and SMITH, D.W.: The XXY syndrome in childhood. Detection and treatment. J. Pediat. *80:* 250 (1972).
13 CASEY, M.D.; SEGALL, L.J.; STREET, D.R.K., and BLANK, C.E.: Sex chromosome

abnormalities in two State hospitals for patients requiring special security. Nature, Lond. *209:* 641–642 (1966).

14 CASEY, M.D.; BLANK, C.E.; STREET, D.R.K.; SEGAL, L.J.; MCDOUGALL, J.H.; MCGRATH, P.J., and SKINNER, J.L.: YY chromosomes and antisocial behaviour. Lancet *ii:* 859–860 (1966).

15 CATTANACH, B.M.; POLLARD, C.E., and HAWKES, S.G.: Sex-reversed mice: XX and X0 males. Cytogenetics *10:* 318–337 (1971).

16 CAUGHEY, J.E. and MYRIANTHOPOULOS, N.C.: Dystrophia myotonica and related disorders (Thomas, Springfield 1963).

17 COURT BROWN, W.M.; HARNDEN, D.G.; JACOBS, P.A.; MACLEAN, N., and MANTLE, D.J.: Abnormalities of the sex chromosome complement in man. Med. Res. Council, Spec. Rept. Ser., No. 305 (Her Majesties Stationery Office, London 1964).

18 COURT BROWN, W.M.; PRICE, W.H., and JACOBS, P.A.: Further information on the identity of 47, XYY males. Br. med. J. *i:* 325–328 (1968).

19 COWIE, J. and KAHN, J.: XYY constitution in a prepubertal child. Br. med. J. *i:* 748–749 (1968).

20 DANKS, D.; ALLAN, J., and ANDERSON, C.M.: A genetic study of fibrocystic disease of the pancreas. Ann. hum. Genet. *28:* 323–356 (1965).

21 DEGROCHY, J.; LAMY, M.; THIEFFREY, S.; ARTHUIS, M. et SALMON, C.: Sysmorphie complexe avec ologophrénie: Délétion des bras courts d'un chromosome 17-18. C.r. hebd. Séanc. Acad. Sci., Paris *256:* 1028 (1963).

22 DE LA CHAPELLE, A.: Nature and origin of males with XX sex chromosomes. Am. J. hum. Genet. *24:* 71–105 (1972).

23 DE LA CHAPELLE, A.; SCHRODER, J.; MURROS, G., and TALLQVIST, G.: Two XX males in one family and additional observations on the etiology of XX males. Clin. Genet. *11:* No. 2 (1977).

24 DE VAAL, O.M.: Genital intersexuality in three brothers, connected with consanguineous marriages in the three previous generations. Acta paediat. *44:* 35–39 (1955).

25 EDWARDS, J.A. and BANNERMAN, R.M.: Familial gynecomastia; in The clinical delineation of birth defects. The endocrine system, pp. 193–195 (Williams & Wilkins, Baltimore 1971).

26 EWER, R.W.: Familial monotropic pituitary gonadotropin insufficiency. J. clin. Endocr. Metab. *28:* 783–788 (1968).

27 FERGUSON-SMITH, M.A.: Karyotype-phenotype correlations in gonadal dysgenesis and their bearing on the pathogenesis of malformations. J. med. Genet. *2:* 142–155 (1965).

28 FERGUSON-SMITH, M.A.: X-Y chromosome interchange in the aetiology of true hermaphroditism and of XX Klinefelter syndrome. Lancet *ii:* 475–476 (1966).

29 GAYRAL, L.; BARRAUD, M.; CARRIE, J. et CANDEBAT, L.: Pseudo-hermaphrodisme à type de «testicule féminisant»: 11 cas. Etude hormonale et étude psychologique. Toulouse méd. *61:* 637–647 (1960).

30 GOLDSTEIN, J.L. and WILSON, J.D.: Hereditary disorders of sexual development in man; in MOTULSKY and LENZ Birth defects, pp. 165–173 (Excerpta Medica, Amsterdam 1974).

31 HALL, J.G.; MORGAN, A., and BLIZZARD, R.M.: Familial congenital anorchia. Birth Defects Original Article Series *11:* 115–119 (1975).

32 Hamerton, J.L.; Dickson, J.M.; Pollard, C.E.; Grieves, S.A., and Short, R.V.: Genetic intersexuality in goats. J. Reprod. Fert. 7: suppl., pp. 25–51 (1969).
33 Hanley, W.B.; McKusick, V.A., and Barranco, F.T.: Osteochondritis dissecans with malformations in two brothers. A review of familial aspects. J. Bone Jt Surg. 49A: 925–937 (1967).
34 Hockaday, T.D.R.: Hypogonadism and life-long anosmia. Post-grad. med. J. 42: 572–574 (1966).
35 Hurxthal, L.M.: Sublingual use of testosterone in 7 cases of hypogonadism: report of 3 congenital eunuchoids occurring in one family. J. clin. Endocr. Metab. 3: 551–556 (1943).
36 Imperato-McGinley, J.; Guerrero, L.; Gautier, T., and Peterson, R.E.: An unusual inherited form of male pseudohermaphroditism. A model of 5-alpha-reductase deficiency in man. Abstract. J. clin. Invest. 53: 35 (1974).
37 Jacobs, P.A.; Price, W.H.; Court Brown, W.M.; Brittain, R.P., and Whatmore, P.B.: Chromosome studies on men in maximum security hospitals. Ann. hum. Genet. 31: 339–358 (1968).
38 Jacobs, P.A.: Structural abnormalities of the sex chromosomes. Br. med. Bull. 25: 94–89 (1969).
39 Kallmann, F.J.; Schoenfeld, W.A., and Barrera, S.E.: The genetic aspects of primary eunuchoidism. Am. J. ment. Defic. 48: 203–236 (1944).
40 Kaplan, E.; Shwachman, H.; Perlmutter, A.D.; Rule, A.; Khaw, K.T., and Holsclaw, D.S.: Reproductive failure in males with cystic fibrosis. New Engl. J. Med. 279: 65–69 (1968).
41 Kasdan, R.; Nankin, H.R.; Troen, P.; Wald, N.; Pan, S., and Yanaihara, T.: Paternal transmission of maleness in XX human beings. New Engl. J. Med. 288: 539–545 (1973).
42 Kaufman, M.; Straisfeld, C., and Pinsky, L.: Male pseudohermaphroditism presumably due to target organ unresponsiveness to androgens. Deficient 5-alpha-dihydrotestosterone binding in cultured skin fibroblasts. J. clin. Invest. 58: 345–350 (1976).
43 Klinefelter, H.F., jr.; Reifenstein, E.C., jr., and Albright, F.: Syndrome characterized gynecomastia, aspermatogenesis without aleydigism, and increased secretion of follicle-stimulating hormone (gynecomastia). J. clin. Endocr. Metab. 2: 615 (1942).
44 Lenz, W.; Nowakowski, H.; Prader, A. und Schirren, O.: Die Ätiologie des Klinefelter-Syndroms. Ein Beitrag zur Chromosomenpathologie beim Menschen. Schweiz. med. Wschr. 89: 727–731 (1959).
45 Le Marquand, H.S.: Congenital hypogonadotrophic hypogonadism in five members of a family, three brothers and two sisters. Proc. R. Soc. Med. 47: 442–446 (1954).
46 Livingstone, F.B.: The distribution of the abnormal hemoglobin genes and their significance for human evolution. Evolution 18: 685–699 (1964).
47 Lubs, H.A., jr.; Vilar, O., and Bergenstal, D.: Familial male pseudohermaphroditism with labial testes and partial feminization: endocrine studies and genetic aspects. J. clin. Endocr. Metab. 19: 1110–1120 (1959).
48 Lyon, M.F. and Hawkes, S.G.: X-linked gene for testicular feminization in the mouse. Nature, Lond. 225: 1217–1219 (1970).
49 MacLean, N.; Mitchell, J.M.; Harnden, D.G.; Williams, J.; Jacobs, P.A.;

BUCKTON, K.E.; PAIKIE, A.G.; COURT BROWN, W.M.; MCBRIDE, J.A.; STRONG, J.A.; CLOSE, H.G., and JONES, C.C.: A survey of sex chromosome abnormalities among 4514 mental defectives. Lancet *i:* 293–296 (1962).

50 MALES, J.L.; TOWNSEND, J.L., and SCHNEIDER, R.A.: Hypogonadtropic hypogonadism with anosmia-Kallmann's syndrome. Archs intern. Med. *131:* 501–508 (1973).

51 MARSHALL, H.K. and HARDER, H.I.: Testicular feminization syndrome in male pseudohermaphrodite: report of two cases in identical twins. Obstet. Gynec., N.Y. *12:* 284–293 (1958).

52 MEYER, W.J., III; MIGEON, B.R., and MIGEON, C.J.: Locus on human X chromosome for dihydrotestosterone receptor and androgen insensitivity. Proc. natn. Acad. Sci. USA *72:* 1469–1472 (1975).

53 MONEY, J.; ANNECILLO, C.; VAN ORMAN, B., and BORGANONKARY, D.S.: Cytogenetics. Hormones, and behaviour disability. Comparison of XYY and XXY syndrome. Clin. Genet. *6:* 351–370 (1974).

54 MOSIER, L.D.; SCOTT, L.W., and COTTER, L.H.: The frequency of the positive sex-chromatin pattern in males with mental deficiency. Pediatrics *25:* 291–297 (1960).

55 MURDOCH, J.L.; WALKER, B.A.; HALL, J.G.; ABBEY, H.; SMITH, K.L., and MCKUSICK, V.A.: Achondroplasia – a genetic and statistical survey. Ann. hum. Genet. *33:* 227–244 (1970).

56 NETTER, A.; LUMBROSO, P.; YANEVA, H. et BELLAISCH, J.: Le testicule féminisant. Annls Endocr. *5:* 994–1014 (1958).

57 NEW, M.K.: Male pseudohermaphroditism due to 17-alpha-hydroxylase deficiency. J. clin. Invest. *49:* 1930–1941 (1970).

58 NIELSEN, J.; FRIEDRICH, U., and ZEUTHEN, E.: Stature and weight in boys with the XYY syndrome. Humangenetik *14:* 66 (1971).

59 NOONAN, J.A.: Hypertelorism with Turner phenotype. A new syndrome with associated congenital heart disease. Am. J. Dis. Child. *116:* 373–380 (1968).

60 OHNO, S. and LYON, M.F.: X-linked testicular feminization in the mouse as a non-inducible regulatory mutation of the Jacob-Monod type. Clin. Genet. *1:* 121–127 (1970).

61 OPITZ, J.M.; SIMPSON, J.L.; SARTO, G.E.; SUMMITT, R.L.; NEW, M., and GERMAN, J.: Pseudovaginal perineoscrotal hypospadias. Clin. Genet. *3:* 1–26 (1971).

62 OVERZIER, C.: The so-called true Klinefelter's syndrome; in OVERZIER Intersexuality, pp. 277–297 (Academic Press, New York 1963).

63 OPPENHEIMER, E.H. and ESTERLY, R.H.: Observations on cystic fibrosis of the pancreas. V. Developmental changes in the male genital system. J. Pediat. *75:* 806–811 (1969).

64 PENROSE, L.S.: Parental age and mutation. Lancet *ii:* 312–313 (1955).

65 PENROSE, L.S. and SMITH, F.G.: Down's anomaly (Little, Brown, Boston 1966).

66 POLAILLON: Observation d'hermaphrodisme. Bull. Soc. Obstet. Gynec., Paris *123:* (1891).

67 QAZI, Q.H.; ARNON, R.G.; PAYDAR, M.H., and MAPA, H.D.: Familial occurrence of Noonan syndrome. Am. J. Dis. Child. *127:* 696–698 (1974).

68 RAO, G.J.S. and NADLER, H.L.: Arginine esterase in cystic fibrosis of the pancreas. Pediat. Res. *8:* 684–686 (1974).

69 REED, M.; SHOKEIR, M.H.K., and MCPHERSON, R.: The hypertelorism-hypospadias syndrome. J. Can. Ass. Radiol. *26:* 240–248 (1975).

70 RICCARDI, V.M.; ATKINS, L., and HOOMES, L.B.: Absent patellae, mild mental retardation, skeletal and genitourinary anomalies, and group C autosomal mosaicism. J. Pediat. 77: 664 (1970).

71 SARTO, G.E.; OPITZ, J.M., and INHORN, S.L.: Considerations of sex chromosome abnormalities in man; in BENIRSCKE Comparative mammalian cytogenetics, pp. 390–413 (Springer, New York 1969).

72 SARTO, G.E. and OPITZ, J.M.: The XY gonadal agenesis syndrome. J. med. Genet. 10: 288–296 (1973).

73 SCHROFFNER, W.G. and FURTH, E.D.: Hypogonadotropic hypogonadism with anosmia (Kallmann's syndrome) unresponsive to clomiphene citrate. J. clin. Endocr. Metab. 31: 267–270 (1970).

74 SHAPIRA, E.; RAO, G.J.S.; WESSEL, H.U., and NADLER, H.L.: Absence of an α_2-macroglobulin-protease complex in cystic fibrosis. Pediat. Res. 10: 812–817 (1976).

75 SHOKEIR, M.H.K.: Autosomal dominant form of hypospadias (in preparation, 1977).

76 SMITH, D.W.; LEMLI, L., and OPITZ, J.M.: A newly recognized syndrome of multiple congenital anomalies. J. Pediat. 64: 210–217 (1964).

77 SPARKES, R.S.; SIMPSON, R.W., and PAULSEN, C.A.: Familial hypogonadotropic hypogonadism with anosmia. Archs intern. Med. 121: 534–538 (1968).

78 SPOCK, A.; HEICK, H.M.C.; CRESS, H., and LOGAN, W.S.: Abnormal serum factor in patients with cystic fibrosis of the pancreas. Pediat. Res. 1: 173–177 (1967).

79 SPITZ, I.M.; DIAMANT, Y.; ROSEN, E.; BELL, J.; BEN-DAVID, M.; POLISHUK, W., and RABINOWITZ, D.: Isolated gonadotropin deficiency. A heterogenous syndrome. New Engl. J. Med. 290: 10–15 (1974).

80 STERNBERG, W.H.; BARCLAY, D.L., and KLOEPFER, H.W.: Familial XY gonadal dysgenesis. New Engl. J. Med. 278: 695–700 (1968).

81 ROSEWATER, S.; GWINUP, G., and HAMWI, G.J.: Familial gynecomastia. Ann. intern. Med. 63: 377–385 (1965).

82 TAYLOR, H.; BARTER, R.H., and JACOBSON, C.B.: Neoplasms of dysgenetic gonads. Am. J. Obstet. Gynec. 96: 816–823 (1966).

83 WACHTEL, S.S.; KOO, G.C.; BREG, W.R.; ELIAS, S.; BOYSE, E.A., and MILLER, O.J.: Expression of H-Y antigen in human males with two Y chromosomes. New Engl. J. Med. 293: 1070–1072 (1975).

84 WACHTEL, S.S.; OHNO, S.; KOO, G.C., and BOYSE, E.A.: Possible role for H-Y antigen in the primary determination of sex. Nature, Lond. 257: 235–236 (1975).

85 WILSON, J.D. and WALKER, J.D.: The conversion of testosterone to 5α-androstan-13β-ol-3-one (dihydrotestosterone) by skin slices of man. J. clin. Invest. 48: 371–379 (1969).

86 WINTERBORN, M.H.; FRANCE, N.E., and RAITI, S.: Incomplete testicular feminization. Archs Dis. Childh. 45: 811–812 (1970).

M.H.K. SHOKEIR, MD, PhD, Division of Medical Genetics, University of Saskatchewan, *Saskatoon, Sask.* (Canada)

The Urologist and Male Infertility

PHILIP G. KLOTZ

Department of Surgery, University of Toronto, and Division of Urology, Mount Sinai Hospital, Toronto, Ont.

The management of the infertile couple involves a number of medical disciplines, hopefully working together. The family physician, the endocrinologist, the psychiatrist, the obstetrician-gynaecologist, the social worker, the urologist, and more recently, the andrologist, all play a role in helping the 15% of couples who are barren. In 30% of these couples, the problem resides primarily in the male partner, and often this is where the urologist plays his role. This chapter will attempt to outline what the relationship of the urologist can be to the problem of male infertility. This involves three main areas – prevention, diagnosis and treatment.

Prevention

Prevention of male infertility is a responsibility of many sectors of the medical community, but involves particularly family physicians and surgeons in the following areas:

Stress. Studies in animals and in men (e.g. astronauts) have shown that factors such as overcrowding [16], altitude [32], and prolonged heat exposure, can reduce spermatogenesis. Emotional stress can also depress sperm production and some schizophrenics have shown testicular degeneration. Stress causes a rise in catecholamine levels, and indeed circulating adrenalin levels are elevated in some infertile males [8] and in patients following prolonged periods of immobilization.

Drugs. Some preparations which may inhibit spermatogenesis are nitrofurantoin [27], anti-metabolites and other drugs used to treat malignancy, anti-malarial compounds, and monoamino acid oxidase antagonists [34], such as pargyline hydrochloride and phenelzine sulphate.

Cryptorchism. Cryptorchism affects 0.3% of males, may be familial, and is associated with chromosomal abnormalities in 60% of cases [23]. Despite treatment, the normally-descended opposite testis may also be defective. Treatment should be undertaken at age 4–5 years. The testis may descend with injections of human chorionic gonadotrophins (HCG), but if not, surgery is necessary.

Testicular torsion. This can result in infarction unless promptly corrected. Surgical exploration should be carried out immediately, correcting the involved side, and fixing the opposite testis to prevent further torsion, since the free-swinging testicle is usually a bilateral condition.

Acute epididymitis. This can result in stenosis of the tiny epididymal duct, or may affect sperm transit time through the epididymis [21, 28]. Prompt treatment with antibiotics, bed rest, and ice packs is essential.

Prostatitis. Prostatitis is considered by many urologists to reduce fertility. The finding of many leucocytes in the semen, and positive semen cultures [14], demands treatment in the infertile male, even though asymptomatic. The possibility that organisms, such as T-mycoplasma and Chlamydia, while non-pathogenic, may affect the prostate and fertility [19, 25], is under scrutiny at present.

Surgical procedures. Retroperitoneal surgery and pelvic surgery may damage the sympathetic nerve supply to the bladder neck, resulting in retrograde ejaculation of semen back into the bladder. This has been reported in patients undergoing para-aortic lymph node dissection for testicular tumour [22], pelvic surgery, and operation to explore the lower ureter. Diabetics are particularly prone to retrograde ejaculation. Transurethral surgery in young males can damage the bladder neck and result in anejaculation, or seal the ejaculatory ducts within the prostate. Spermatocoele excision should be postponed until after a man has had children, as this operation may damage the epididymis and result in a sterile testicle. It is important that family physicians and surgeons be aware of the possible effect of such conditions on male fertility, and plan their treatments with this in mind.

Diagnosis

As in all other fields of medicine, accurate diagnosis in male infertility points the way to appropriate treatment. It may also indicate that no possible treatment is currently available for certain males, suggesting that the couple proceed directly to artificial insemination or adoption, if these routes are desired. Unfortunately, these options are often necessary, but hopefully new forms of therapy will emerge as knowledge expands. A step-by-step approach is indicated, and may require the services of the endocrinologist, the psychiatrist, and the social worker. This chapter will outline the role of the urologist in this area.

History

Childhood illnesses such as mumps, trauma, cryptorchism, and delayed puberty are important. Sexual function should be specifically inquired after, especially erection, penetration, orgasm and ejaculation. The frequency of intercourse (or its infrequency, e.g. once a month) may indicate the reason for the infertility. A history of previous marriage, and the pregnancy, or lack of it, that resulted, may shed light on the problem. The general health and a history of previous venereal disease, diabetes, mental illness, or drugs taken recently, should be brought out. Exposure to temperature extremes and the use of tight underwear, may suggest a reason for his problem.

Physical examination

This should be thorough and must include the presence or absence of secondary sex characteristics, the blood pressure, and the state of the genitalia. The most important feature is the size and consistency of the testes. Familiarity with these parameters comes with experience. Often, the infertile testis is smaller and softer than normal.

The size and normality of the penis are important when hypospadias is present, as other genital abnormalities occur in 5% of cases. Varicocoele should be checked for in a good light, with the patient standing and performing Valsalva's manoeuvre. Varicocoeles usually are left-sided, but may be bilateral. Their possible role in subfertility will be discussed later.

Semen Analysis

This is one of the most essential steps in diagnosis. The specimen should be obtained by masturbation directly into a clean, dry glass bottle, following 3 days of sexual abstinence. Plastic containers may affect sperm motility.

Table I. Seminal fluid analysis

Volume	1.5–3.5 cm^3
Lysis	to smooth consistency within 10 min of ejaculation
Viscosity	homogeneous and fluid
Sperm count	>20 million/cm^3
Sperm motility	>60% forward motion
Staining	>75% normal forms

At least two specimens should be examined and within 2 h of production.

While the number of studies which can be done is large, table I indicates the most clinically significant factors.

Haematology, biochemical tests of serum and urine, and X-ray studies all form part of the investigation of the infertile male, depending on the specific problems involved, and should be used where indicated. For the urologist however, the investigation of the infertile male will, by this stage, have put the patient into one of the following categories:

(1) Apparently normal fertility: i.e. no gross abnormatity detected on history and physical, with normal semen analysis. Such patients may still have a problem with fertility associated with male-female antigen-antibody reactions, or other rare factors, but at the moment this will not be apparent.

(2) Oligospermia: less than the normal number of spermatozoa, or increased numbers of abnormal or non-motile spermatozoa, and resultant impaired fertility.

(3) Azoospermia: no sperm in the ejaculate, and sterility as a result.

At this point in the investigation, it is wise to see both partners in consultation, outline the problem, and discuss the future programme they may embark upon, if they wish. They may decide to go no further, and adopt or seek artificial insemination. They should have the possibilities, probable success rates, and possible complications of further investigation and resultant treatment fully explained to them. Should the decision to continue be taken, the urological investigation proceeds as follows.

Testicular Biopsy

The prime indication for biopsy of the testes is azoospermia, to determine if spermatozoa are being produced in the testes. Biopsy in patients with oligospermia is rarely of any practical value, although it may show a variety of abnormalities of sperm maturation. In azoospermia, biopsy is often accompanied by vasography. Biopsy can be useful in endocrine disturbances

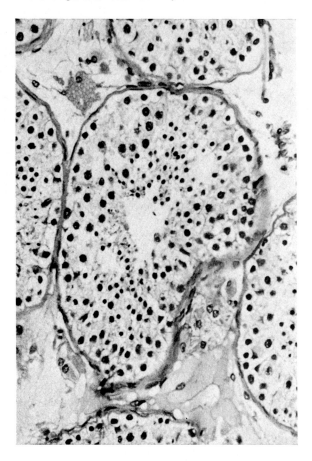

Fig. 1. Biopsy of normal testis.

to differentiate primary from secondary hypogonadism. As a research tool, biopsy is important in trying to relate abnormalities in semen, to structural changes in the seminiferous cells, but this is not of much value to the immediate problem of the infertile couple (fig. 1–3).

It should be borne in mind that testicular biopsy may depress normal spermatogenesis for up to 4 months, and occasionally may cause permanent damage to the testes [2].

The technique is quite simple and may be done under general anaesthesia, or following local anaesthetic infiltration of the spermatic cord and the overlying scrotal skin. A transverse incision 1–2 cm long in the scrotum and

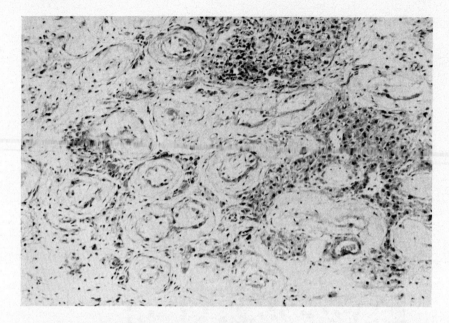

Fig. 2. Testicular atrophy. Note absence of progressive spermatogenesis, small tubules, thickened basement membrane.

tunica albuginea is made and a small portion of seminiferous tissue removed. It should be placed immediately in Bouin's solution, which does not distort the microscopic anatomy, as will formalin. The tunica and scrotal skin are then closed with one or two absorbable sutures.

Seminal Vesiculography and Vasography

The prime indication for this procedure is the patient with azoospermia and normal spermatogenesis on testicular biopsy. Such an individual may have a bilateral conductive block and may be helped surgically. Other indications for this study are not related to infertility, but to give information on infection of the seminal vesicles, tumours of the vesicle, and trauma.

The technique can be done by either of two routes, and is often carried out at the same time as biopsy. The first method is to put a small catheter into the ejaculatory duct via a panendoscope, and inject radio-opaque solution. In the second method, the vas can be exposed in the scrotum, close to the testis, and cannulated with a blunt No. 25 or No. 26 needle. Dye is then injected and X-rays taken. Some experts rely on the injection of saline to

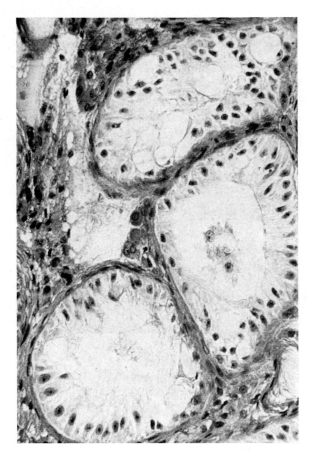

Fig. 3. Sertoli-cell-only syndrome. 'Picket-fence' appearance of tubules.

detect increased resistance at a block in the vas, but radio-opaque dye is much more revealing and accurately delineates the vas and seminal vesicle (fig. 4). The epididymis can be outlined by injecting the dye in the other direction, towards the testicle, but these studies are usually difficult to interpret and of little value at present.

The combination of azoospermia, normal spermatogenesis on testicular biopsy, and a patent vas and ejaculatory duct on vasography, means that the male has blockage of both epididymal ducts. Such obstruction may be congenital or post-inflammatory due to previous gonorrhoea or non-specific infection. Such patients are candidates for epididymo-vasostomy.

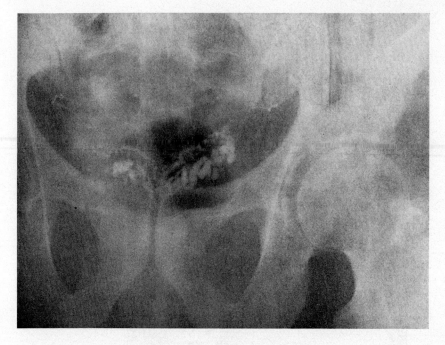

Fig. 4. Normal vasogram and vesiculogram, bilateral.

Treatment of Male Infertility

Apparently Normal Fertility

It is not unusual to see an infertile couple in whom no abnormality can be detected in either partner. The problem may be due to many factors, some of which are known, and many others of which hopefully will be revealed as research progresses. The problem may be associated with abnormalities of the cervical mucus, with sperm antigen-antibody reactions, with difficulties in sexual function, or clothing. The couple should be instructed in sexual technique and informed of the probable time of ovulation. Intercourse should take place at 48-hour intervals to allow for sperm regeneration by the testes. The penis should be left in the vagina after ejaculation until it either slips out spontaneously, or after 10 min, to act as a plug to prevent the escape of semen. Some urologists suggest that the female partner assume the knee-chest position after ejaculation, again to prevent loss of semen from the vagina and allow a pool of semen to collect over the cervical os. The use of tight underwear, such as jockey shorts, should be abandoned for loose-fitting underclothing, as testicular temperature may be elevated, with impaired sperm production and function.

Occasionally, a holiday from the stresses and strains of life may be beneficial, in a non-specific way.

Such cases of apparently normal fertility can be very frustrating, when success seems so close and yet pregnancy is elusive.

Treatment of Oligospermia

The measures suggested for the apparently fertile male also apply to those with oligospermia. If any underlying endocrine or metabolic problem can be detected, then it should be corrected. Many empiric modalities of treatment have been tried over the years to improve semen quality, with occasional success [33]. The pharmacological approach to oligospermia will be discussed in a subsequent chapter.

The following therapeutic approaches may be of some value in appropriate cases.

Coitus interruptus. In most oligospermic men the first portion of the ejaculate has a higher sperm count than the last portion. This can be documented by split ejaculate studies [15]. In addition, it is thought that the seminal vesicle secretion, which is added to the last portion of the ejaculate, may impair fertility in some males.

To utilize this knowledge [30], the male is instructed to withdraw from the vagina as soon as the first spurt of ejaculation occurs. This technique has had some degree of success [10].

Sub-clinical non-specific prostatitis. This is considered by some experts to depress sperm function. Certainly, prostatic secretions are necessary for normal sperm motility. The possibility that the products of infection in the prostate can influence fertility, is very attractive. The prostatic fluid should be examined microscopically for leucocytes, which normally number less than 20/hpf. Cultures of the prostatic fluid may be significant, but the problem lies with contamination as the fluid progresses down the urethra, which normally contains bacteria in its distal 2–3 cm. It has been suggested that organisms such as Chlamydia and T-mycoplasma are capable of impairing sperm function, but this has not as yet been proven. It would appear wise, when prostatitis is suspected, to treat both partners with a course of antibiotics, even though clinically there are no symptoms.

Table II. Results of varicocoelectomy

Author	Year	Cases	Improvement in semen		Pregnancies	
			n	%	n	%
FRITJOFFSON and ABREN [18]	1967	40		62.5		17
CHARNY and BAUM [7]	1968	104		64		24
BROWN [5]	1968	117		not given		40
DUBIN and HOTCHKISS [12]	1969	88		75		33
DUBIN and AMELAR [11]	1975	504	361	71	276	55
STEWART and MONTIE [33]	1973	20	17	70	11	55

Varicocoelectomy has become the main line of attack on oligospermia [11]. Varicocoele is present in 30–40% of men, usually on the left side. While many men with varicocoele have produced progeny, the results of high ligation for varicocoele in oligospermic infertility appear to be highly significant (table II). Theoretically, varicocoele may impair spermatogenesis: (1) by interfering with heat exchange in the testis, which is normally cooler than corporeal temperature by 2 °C; (2) by altering the hydrodynamics of the testis by retrograde blood flow in the spermatic vein; (3) by delivering higher quantities of adrenal gland products to the left testis, since both left adrenal and left spermatic veins drain into the left renal vein. Measurements of cortisol have been found to be higher in the left spermatic vein than in the systemic venous circulation [4].

Venography has shown cross-anastomosis of the venous drainage between left and right testes via the abdominal wall venous plexuses, suggesting a bilateral effect of unilateral varicocoele [17].

Studies of seminal fluid have shown a 'stress pattern' in patients with varicocoele [24]. The spermatozoa have decreased motility and there is an increase in immature forms of spermatozoa. Usually the sperm count is lowered, but not necessarily.

The results of surgery are better when the preoperative sperm count is higher than 10 million/cm^3. Above this range (table II) semen quality improvement occurred in 60–70%, and the pregnancy rate was 17–55%, depending on the series [13]. The improved results are unrelated to the size of the varicocoele and small varicocoeles should be searched for diligently. Most pregnancies occur within 6 months of surgery, but can be delayed as long as 1½ years.

Table III. Results of epididymo-vasostomy

Author	Year	Cases	Sperm in ejaculate		Pregnancies	
			n	%	n	%
Hagner [20]	1936	65	21	31	16	25
Michelson [26]	1947	8	3	38	2	25
Bayle [3]	1950	70	33	47	21	30
O'Conor [29]	1953	61	14	23	5	8
Hanley and Hodges [21]	1959	13	13	100	not given	

Treatment of Azoospermia

Productive azoospermia. The failure of the testes to produce spermatozoa is the field of the endocrinologist and outside the scope of this paper. A variety of conditions, Klinefelter's syndrome, maturation arrest, and Sertoli-cell-only syndrome may account for some of the cases of inability to produce sperm.

Conductive azoospermia. This occurs when there is sperm production by the testes, but bilateral blockage in the system of delivery of sperm to the ejaculate. This may be present anywhere from the rete testis to the ejaculatory ducts in the prostate. The commonest cause today is a previous vasectomy in a man who now desires children by a second wife.

Millions of vasectomies are being done throughout the world. In the US it is considered that 1:400 to 1:500 vasectomized males may later request reversal. Many techniques are being assessed in implanting reversible vasectomy devices, both rigid and soft in type, with no consistent results as yet [6]. Present techniques of vas re-anastomosis are quite successful, such that sperm reappears in the ejaculate in 80–90% of cases [9]. However, the pregnancy rate remains at 35–50%, because the prolonged obstruction produced by vasectomy results in some permanent damage to the seminiferous cells of the testis, which may be largely irreversible [31].

Obstruction in the epididymis can occasionally be relieved by epididymo-vasostomy, which short-circuits the epididymis.

A side-to-side anastomosis is made between the epididymis and vas deferens with fine sutures, in the hope of by-passing an epididymal obstruction. The results of this operation in some hands are shown in table III.

One of the problems with this procedure, aside from the high incidence of occlusion of the anastomosis, is the problem of sperm maturation as it progresses along the epididymis. In the short-circuiting procedure, this process does not take place, resulting often in poor quality spermatozoa in the ejaculate, if they appear at all.

Attempts to produce an artificial spermatocoele [35] using saphenous vein grafts to vas or epididymis have to date been unsuccessful. Hopefully, further efforts will be made to allow a pool of spermatozoa to collect and be aspirated for artificial insemination into the donor's wife.

The problem of retrograde ejaculation of semen back into the bladder can occasionally be overcome by plastic procedures at the bladder neck [1]. If not, some success has been reported in retrieval of the semen from the bladder by catheter, following ejaculation, and using the aspirate for artificial insemination.

Summary

The urologist has an important role to play in the prevention, diagnosis and treatment of male infertility. The accurate diagnosis of the cause of infertility gives the infertile couple an idea of whether to continue with treatment, consider adoption or artificial insemination, or to abandon plans for a family. The parameters of treatment available are outlined, with the realization that our present knowledge of infertility and its correction merely represents the tip of the iceberg. Future developments, spurred by the present intense interest in infertility, are awaited with interest.

References

1   ABRAHAMS, J.; SOLISH, G.I.; BOORJIAN, P., and WATERHOUSE, R.K.: Surgical correction of retrograde ejaculation. J. Urol. *114:* 888 (1975).
2   BAYLE, H.: Masculine sterility. Urol. cutan. Rev. *54:* 129 (1950).
3   BAYLE, H.: Stérilité masculine. Statistique de 95 cas d'azoospermie par oblitération explorés chirurgicalement et traités par l'anastomose épididymo-déférentielle. Presse méd. *58:* 276 (1950).
4   BROWN, J.S.: Varicocoelectomy in the sub-fertile male: a ten year experience with 295 cases. J. Fert. Steril. *27:* 1046 (1976).
5   BROWN, J.S.: Varicocoelectomy in sub-fertile men. Gen. Pract. *94:* 37 (1968).
6   BRUESCHKE, E.E., *et al.:* Reversible vas occlusion device. J. Fert. Steril. *26:* 29 (1975).

7   Charny, C.W. and Baum, S.: Varicocoele and infertility. J. Am. med. Ass. *204:* 1165 (1968).
8   Combaire, F. and Vermeulen, A.: Varicocoele sterility. Cortisol and catecholamines. J. Fert. Steril. *25:* 28 (1974).
9   Dorsey, J.W.: Surgical correction of post-vasectomy sterility. J. Urol. *110:* 554 (1973).
10 Dubin, L.: Discussion at W.W. Scott Symposium on Male Infertility, Rochester 1976.
11 Dubin, L. and Amelar, R.D.: Varicocoelectomy as therapy in male infertility: a study of 504 cases. J. Fert. Steril. *26:* 217 (1975).
12 Dubin, L. and Hotchkiss, R.S.: Testis biopsy in sub-fertile men with varicocoele. J. Fert. Steril. *20:* 50 (1969).
13 Dubin, L. and Amelar, R.D.: Paper presented at W.W. Scott Symposium on Male Infertility, Rochester 1976.
14 Elliason, R.: Paper presented at W.W. Scott Symposium on Male Infertility, Rochester 1976.
15 Elliason, R.: Standards for investigation of human semen. Andrologie *4:* 49 (1971).
16 Ellis, L.C.: Inhibition of rat testicular androgen synthesis *in vitro* by melatonin and serotonin. Endocrinology *90:* 17 (1972).
17 Eiriby et al.: Sub-fertility and varicocoele. Venogram demonstration of anastomosis sites in sub-fertile men. J. Fert. Steril. *26:* 1013 (1975).
18 Fritjoffson, A. and Abren, C.: Studies on varicocoele and sub-fertility. Scan. J. Urol. Nephrol. *1:* 55 (1967).
19 Gnarpe, H. and Friberg, J.: T-mycoplasma as a possible cause for reproductive failure. Nature, Lond. *242:* 120 (1973).
20 Hagner, F.R.: The operative treatment of sterility in the male. J. Am. med. Ass. *107:* 1851 (1936).
21 Hanley, G. and Hodges, R.D.: The epididymis in male sterility: a preliminary report of microdissection studies. J. Urol. *82:* 508 (1959).
22 Kedra, K.; Markland, C., and Fraley, E.: Sexual function following high retroperitoneal lymphadenectomy. J. Urol. *114:* 237 (1975).
23 Lipshultz, L.I.: Cryptorchism and the infertile male. J. Fert. Steril. *27:* 609 (1976).
24 MacLeod, J.: Seminal cytology in presence of varicocoele. J. Fert. Steril. *16:* 735 (1965).
25 Matthews, C.D.; Elmslie, R.G.; Clapp, K.H., and Surgios, J.M.: The frequency of genital mycoplasma infection in human fertility. J. Fert. Steril. *26:* 743 (1975).
26 Michelson, L.: Treatment of azoospermia by vaso-epididymal anastomosis. West. J. Obstet. Gynec. *55:* 120 (1947).
27 Nelson, W.O. and Bunge, R.G.: The effect of therapeutic doses of nitrofurantoin (Furadantin) upon spermatogenesis in man. J. Urol. *77:* 275 (1957).
28 Nilsson, S.; Obrant, K.O., and Persson, P.S.: Changes in the testis parenchyma caused by acute non-specific epididymitis. J. Fert. Steril. *19:* 748 (1968).
29 O'Conor, V.J.: Mechanical aspects of surgical correction of male sterility. J. Fert. Steril. *4:* 439 (1953).
30 Phadke, A.M.; Naragan, R.S., and Shubhada, D.D.: Seminal fructose content in necrospermia. J. Fert. Steril. *26:* 1021 (1975).
31 Plant, S.M.: Testicular morphology in rats vasectomized as adults. Science *181:* 554 (1973).

32 SEGAL, S.; SADOVSKY, E.; PALTI, Z.; PFEIFER, Y., and POLISHUK, W.Z.: Serotonin and 5-HIAA in fertile and sub-fertile men. J. Fert. Steril. *26:* 314 (1975).
33 STEWART, B.H. and MONTIE, J.E.: Male infertility: an optimistic report. J. Urol. *110:* 216 (1973).
34 URRY, R.L. and DOUGHERTY, M.S.: Inhibition of rat spermatogenesis and seminiferous tubule growth after short-term and long-term administration of a monoamine oxidase inhibitor. J. Fert. Steril. *26:* 232 (1975).
35 VICKERS, M.A.: Creation and use of a scrotal sperm bank in aplasia of vas deferens. J. Urol. *114:* 242 (1975).
36 WONG, T.; STRAUSS, F.H., and WARNER, N.E.: Testicular biopsy in the study of male infertility. Archs Path. *95:* 151 (1973).

P.G. KLOTZ, MD, FACS, FRCS (C), Assistant Professor of Surgery, University of Toronto, *Toronto, Ont.* (Canada)

Use of Diagnostic Ultrasound in the Evaluation of Testicular Disorders

Murray Miskin and Jerald Bain

Departments of Radiology, Medicine and Obstetrics and Gynecology,
University of Toronto and Mount Sinai Hospital, Toronto, Ont.

Introduction

Ultrasonography of the testes is a rapid, painless, harmless [4, 5] investigative technique. It allows the examiner to assess testicular size and internal consistency. Particularly when scrotal swelling precludes adequate physical examination, ultrasound provides a method of 'seeing' into the scrotal sac to help determine the nature of the abnormality.

Diagnostic ultrasound employs high frequency sound waves [2.5–10 million cycles/sec (2.5–10 MHz)] to provide images of various parts of the body. The technique is based on the physical phenomenon that different tissues have different acoustic properties so that a sound beam passing from one tissue into another will be transmitted, refracted, or reflected at the interface between the two tissues. Those waves reflected back toward the original source are processed by the ultrasound equipment to produce an image of the part being examined. The source of the sound waves is a transducer which is moved over the skin of the patient. The transducer is a small porcelain cylinder which contains a piezoelectric crystal. This type of crystal vibrates and produces sound waves when it is shocked by a current of electricity. Conversely, when it is agitated by returning echoes, it produces an electrical current. The current is processed by the ultrasound equipment and an image of the echo-producing structure in the body is made. The ultrasonic waves are pulsed so that only one transducer is necessary to produce the sound waves and to listen for returning echoes between the pulses. The skin over which the transducer moves must be coated with a thin layer of oil (vegetable or mineral) or a gel, since ultrasonic waves are reflected by even a very thin layer of air. A coupling agent such as oil enables sound waves to be transmitted from the transducer into the body.

Technique

Examinations may be performed with any standard B-mode ultrasonic apparatus. Most of the instruments on the market today employ Grey scale. This modification of the conventional bi-stable scanner provides more information about the texture of the tissues being examined. While quite a bit of information can be obtained with conventional B scans, Grey scale sonography is the method of choice.

The ultrasonic examination is not difficult to perform. The patient is placed supine on a stretcher and the scrotum is coated with the coupling agent. One gonad is examined at a time. The examiner immobilizes the testis with his left hand and moves the transducer slowly and gingerly from the cranial to caudal pole all the while observing the image being built up on an oscilloscope or television screen. An assistant is required to manipulate the controls on the equipment because both the examiner's hands are occupied making the scans. At least three longitudinal sonograms through each testis should be made – in neutral position, internal and external rotation. Because the testis is an oval, three-dimensional structure, scanning in these three planes will ensure that the sound waves traverse virtually all of the testicular structure. It is not practical to perform transverse sonograms because the testis cannot be immobilized well enough to obtain reproducible echoes.

A transducer of at least 5 MHz should be used in these studies. The thickness of tissue which has to be traversed by the sound waves is not very great, so that the 5 MHz transducer affords sufficient penetration. Finer detail can be appreciated with this transducer than can be obtained with one of lower frequency.

Permanent records of completed sonograms can be made using polaroid, 70 mm film or an 8×10 radiographic film. (The latter method is a fairly recent innovation which employs a camera mounted in a box at the other end of which is a television monitor. Nine separate images can be recorded on each film which can easily be processed in any conventional X-ray processor.)

Normal Testis

The normal adult testis is approximately 2 cm thick and 4 cm long. Ultrasonically, it is visualized as a discrete oval. With conventional bi-

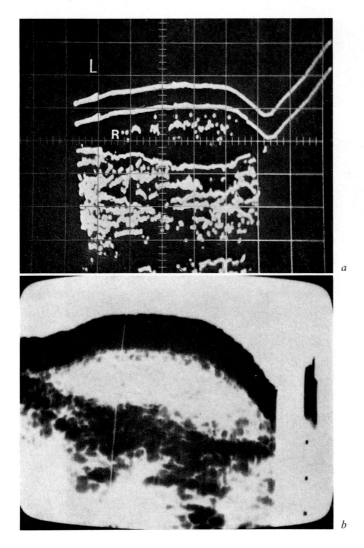

Fig. 1. a Bi-stable ultrasonogram of a normal adult testis. Note the discrete oval outline and the band of echoes in the superficial portion originating from the rete testis (R). Each square on the sonogram represents an area 1 × 1 cm. *b* Grey scale ultrasonogram of a normal testis showing a homogeneous faint echo pattern and a clearly defined testicular outline.

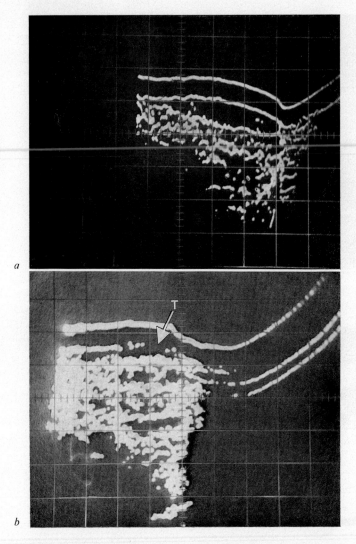

Fig. 2. a A small testis in a patient with oligospermia is shown. Note the small size of this testis relative to the normal testes shown in figure 1. *b* A patient with Klinefelter's syndrome had a testis (T) barely discernible with ultrasonography.

stable scanning the testis is relatively echo-free except for a thin band of echoes appearing in its superficial portion (fig. 1a, b). These are thought to arise from the rete testis [1]. With Grey scale, the testis shows a homogeneous faint echo pattern and inexplicably the rete cannot be distinguished.

The size of the testis is easily measured using a row of dots, each placed 1 cm apart, which may be superimposed on the television or oscilloscope screen, by means of electronic calipers with which most ultrasound equipment is provided or by using a grid superimposed on an oscilloscope screen. An example of a small testis is illustrated in figure 2a and the barely discernible testis of a patient with Klinefelter's syndrome is shown in figure 2b.

The presence of a mass within the testis can be appreciated by the appearance of echoes within the otherwise homogeneous testicular structure. A disruption of a normal testicular outline may also be clearly seen as may fluid surrounding the testis.

Abnormalities

Neoplasms

The most common cause for testicular enlargement is neoplasm. Neoplastic involvement may be localized or diffuse (fig. 3a–c). Diffuse infiltration of the testis produces a large organ with many internal echoes in an irregular pattern, while a localized neoplasm is seen as a cluster of echoes in an otherwise homogeneous structure. It is not yet possible to differentiate types of testicular tumors on the basis of their ultrasound appearance. It is also not possible to distinguish between orchitis and a testis diffusely infiltrated with a tumor.

Testicular Abscess and Hematoma

Testicular abscess produces a diffusely enlarged testis. There is an increase in the number of internal echoes with interspersed echo-free areas. The latter are likely formed by edema whereas the former result from areas of necrosis. The testicular outline becomes indistinct, probably due to inflammatory infiltration of surrounding tissues (fig. 4).

Testicular hematoma presents an appearance similar to abscess. The areas of sonolucency seen in a ruptured testis results from hemorrhage. The testis loses its discrete outline because its tunic has been breached secondary to the trauma.

Varicocele

Varicoceles cannot usually be appreciated clinically when the patient is recumbent. Similarly, a varicocele cannot be defined ultrasonically while the patient is lying down. While it is a little awkward to perform an ultra-

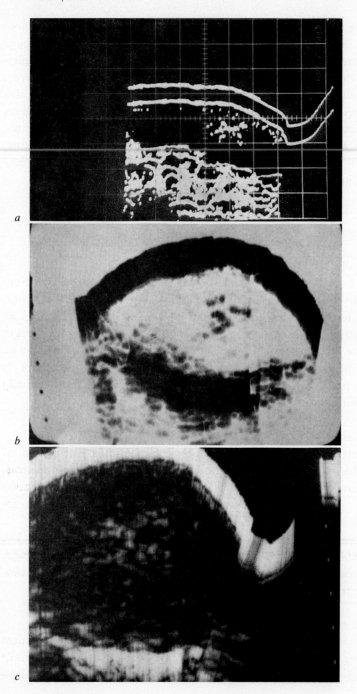

Use of Diagnostic Ultrasound in the Evaluation of Testicular Disorders 123

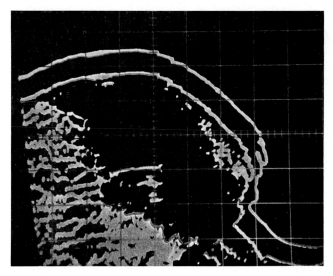

Fig. 4. Testicular abscess. The testis is enlarged, has an irregular deep border, and contains large areas of sonolucency. The latter is most likely due to edema.

sound study while the patient is sitting, this can be accomplished. The veins can be made to stand out even more if the patient is asked to perform the Valsalva maneuver while the sonogram is being performed. A varicocele is visualized as a cluster of tortuous echoes in the deep part of the testis. These are more obvious toward the cranial pole (fig. 5).

Hydrocele

While a hydrocele can be very easily diagnosed clinically, ultrasonography provides a visual record of the condition. Moreover, the hydrocele sac precludes palpation of the underlying testis and assessment of its status. With diagnostic ultrasound, such a testis can be clearly visualized. The hydrocele sac appears as a zone of sonolucency surrounding the testis. No

---

Fig. 3. a An embryonal cell carcinoma is seen as a localized cluster of echoes in the right testis of a 21-year-old man. *b* A seminoma involving this testis is also seen as a cluster of echoes in a testis which is slightly enlarged measuring just over 3 cm in width. *c* Note the marked enlargement of this testis (more than 5 cm in thickness) which is diffusely infiltrated by an embryonal cell carcinoma. There is an increased number of echoes in it and these have an irregular distribution.

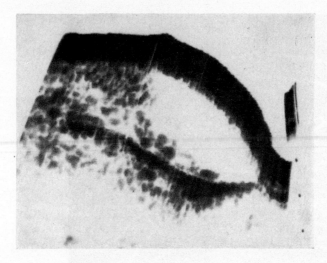

Fig. 5. Varicocele. Note the worm-like cluster of echoes in the deep part of this testis. These arise from the tortuous veins which fill when the patient is in the upright position. This sonogram was made with the patient sitting and performing the Valsalva maneuver. Reproduced by permission from the *Journal of Urology*.

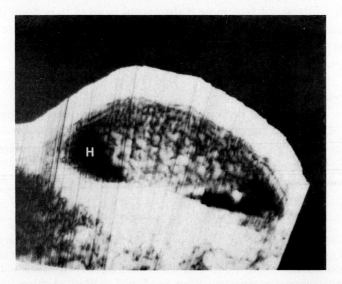

Fig. 6. Hydrocele. Note the sonolucent areas at either pole of the testis. These represent fluid in a hydrocele sac (H).

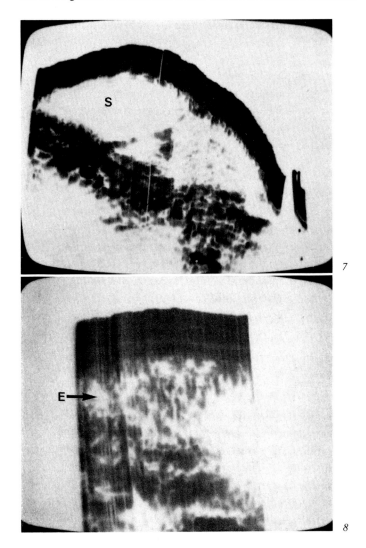

Fig. 7. A spermatocele (S) is visualized as an echo-free area at the cranial pole of the testis.

Fig. 8. Epididymitis. The inflamed epididymis (E) presents as a mass containing many echoes which is visualized at the cranial pole of the testis. The figure is a coned-down sonogram of the region of the cranial pole of the affected testis.

echoes are present in the sac because there are no tissue interfaces to reflect the sound (fig. 6).

Epididymitis and Spermatocele

Both these abnormalities are seen at the upper pole of the testis. A spermatocele is visualized as a clear space because it contains a fluid (fig. 7). The inflamed epididymis is visualized as a solid mass (containing echoes) which appears separate from the testis caudal to it (fig. 8).

Testicular Width Correlated with Sperm Count

The width of each testis can be measured by ultrasonography as indicated above. Because of the difficulty of immobilizing the testis, testicular length cannot be easily assessed by this technique. In a group of 11 men, each of whom had average sperm counts in excess of 50 million/ml of ejaculate, the mean echographic width of the right testis was 2.25 cm, of the left 2.21 cm and of both testes combined 2.31 cm. Lower sperm counts were associated with decreased testicular width. This is illustrated in table I which demonstrates that as sperm count falls the width of each testis also diminishes. This finding is not unexpected since the seminiferous tubules within which spermatogenesis takes place comprise the major component of testicular volume.

Table II shows the mean testicular width of both testes taken together within each sperm count group in a total of 53 men. When the sperm count is less than 30 million/ml mean testicular width is significantly smaller than when the sperm count is greater than 50 million/ml.

The converse can also be demonstrated and this is seen in table III. When testicular width categories are considered alone the sperm counts of all individuals within those categories can be seen to increase with increasing testicular width. The size of the testes bears a direct relationship to the number of sperm produced.

Clinical and Echographic Testicular Widths Compared

The width of each testis in 56 men was measured both ultrasonographically and clinically. The clinical measurement was performed by immobilizing each testis against its corresponding thigh and determining the distance

Table I. Mean echographic testicular width of each individual testis correlated with sperm count group

Sperm count group $\times 10^6$/ml	Mean width right testis, cm	Mean width left testis, cm
>50	2.25	2.21
>40–50	2.30	1.98
>30–40	1.93	1.76*
>20–30	1.93*	1.83*
>10–20	1.89*	1.85*
> 0–10	1.78*	1.79*
0	1.58*	1.52*

* Significantly different from $>50 \times 10^6$/ml group.

Table II. Mean echographic testicular width correlated with sperm count group

Sperm count group $\times 10^6$/ml	Number	Mean testicular width, cm
>50	11	2.23
>40–50	5	2.14
>30–40	3	1.85
>20–30	8	1.88*
>10–20	9	1.86*
> 0–10	12	1.79*
0	5	1.55*

* Significantly different from testicular width at sperm count $>50 \times 10^6$/ml.

Table III. Mean sperm count within each testicular width group measured by echography

Testicular width group (both testes combined) cm	Mean width for each group cm	Number	Sperm count $\times 10^6$/ml
1.7	1.41	15	9.6
1.7–2.0	1.87	16	34.6
2.1–2.4	2.24	15	83.9
2.4	2.59	5	195.0

Correlation coefficient = 0.933.

Table IV. Comparison of mean testicular width of 56 men by sonography and clinical assessment

	Mean testicular width, cm		Correlation coefficient
	sonography	clinical	
Right	1.99	2.67	0.706
Left	1.92	2.58	0.622
Combined	1.96	2.63	0.682

between the immobilizing fingers, the index finger and thumb. Table IV shows the mean results for all the men for each testis and all the testes combined. Manual measurement is consistently slightly larger than that achieved by ultrasound. This can be accounted for by the skin and subcutaneous tissue included in the clinical measurement as well as the increased difficulty in measuring the testis directly rather than from the echographic picture. It can be seen that the right testis, by either measurement technique is slightly larger than the left.

Ultrasound as a Method of Male Contraception

Diagnostic ultrasound is a technique that employs high frequency sound waves. Sound waves are mechanical vibrations which possess energy. As these waves progress through tissue, their energy is dissipated as heat. It can be appreciated, therefore, that any biological tissue exposed to sound waves may, depending on the energy of the sound, experience a measurable rise in temperature. Because of the compressions and rarefactions which constitute a sound wave, such tissue may also experience the phenomenon of cavitation. This refers to the formation of microbubbles from dissolved gases during the negative pressure phase of the sound wave [3]. If either of these phenomena – heat or cavitation – are intense enough, tissue damage may result.

Exposure of mammalian testes to elevated temperatures has a deleterious effect on spermatogenesis [2, 6, 7]. The question arises then, as to whether or not diagnostic ultrasound might affect spermatogenesis either by raising the temperature of the testis being examined or by causing cavitation in its tissues. The answers to these questions hinge on the energy levels used and the duration of exposure of the insonated tissue. The lowest threshold for biological damage is of the order of 11 W/cm² of surface of the crystal generating the sound waves [8]. In diagnostic ultrasound, energy levels of 1–50 μW/cm² are used. Except for Doppler ultrasound, which employs continuous sound waves, diagnostic ultrasound instruments use intermittent short pulses of sound waves. There may be up to 1,000 pulses/sec with an average duration of 1 μsec. The actual time a transducer is in contact with the skin of a patient during an average ultrasound study is in the order of 5 min. These figures, therefore, illustrate that the actual time to which tissues are exposed to ultrasound energy and the energy levels themselves are very low. Indeed, much work has been done to see if any deleterious affects on tissues exposed to diagnostic ultrasound could be demonstrated and if any genetic changes could be produced in laboratory animals. To date, no such adverse affects have been produced when the experimenter employed equipment which produced the same energy levels as ultrasound instruments currently in clinical use.

Recently, there has been work published on the use of ultrasound as a male contraceptive. FAHIM et al. [1] have shown that animal and human testes exposed to 1–2 W/cm² of pulsed ultrasound for periods of 5 min exhibit decreased spermatogenesis. Their work on humans was done on men who later underwent orchiectomy for carcinoma of the prostate. These and the animal testes were examined pathologically and in some cases demonstrated partial degranulation of seminiferous tubules. The animals were found to be hypofertile up to 10 months depending on the dosage, duration, and frequency of exposure to ultrasound. While this work is extremely interesting with respect to the potential use of ultrasound as a male contraceptive, these results cannot be extrapolated to the use of diagnostic ultrasound in the examination of the human testes. The energy levels used in this technique are a mere fraction of that used in the experiments of FAHIM et al. [1] and the actual exposure to the sound waves is for a much shorter period to time.

Conclusions

Ultrasonography of the testes is a new tool in the diagnosis of testicular disorders. More experience with this technique will be needed to determine its ultimate usefulness. It is clear, however, that certain intratesticular lesions can be visualized and differentiated. Diagnostic ultrasound is a noninvasive technique that is simple to perform, safe, inexpensive, and lends itself to ready interpretation. It would appear that in some instances ultrasonography may either suggest that surgery is necessary or that a more conservative approach to therapy can be carried out.

Ultrasonography has no clear-cut advantage over clinical assessment in determining testicular width, although the width as calculated by echography is undoubtedly more accurate. This study has clearly demonstrated, however, that testicular width is directly related to sperm production.

The usefulness of ultrasonography as a safe reversible form of male contraception has yet to be demonstrated.

References

1. FAHIM, M.S.; FAHIM, Z.; DER, R.; HALL, D.G., and HARMAN, J.: Heat in male contraception (hot water 60°C, infrared, microwave, and ultrasound). Proc. FASEB, vol. II, No. 5, pp. 549–562 (1975).
2. FUKUI, N.: Action of body temperature on the testicle. Jap. med. Wld. *3:* 160–163 (1923).
3. KING, D.L. and LELE, P.P.: Biologic effects of diagnostic ultrasound; in KING Diagnostic ultrasound, pp. 290–298 (Mosby, St. Louis 1974).
4. MISKIN, M. and BAIN, J.: B-mode ultrasonic examination of the testes. J. clin. Ultrasound *2:* 307–311 (1974).
5. MISKIN, M.; BUCKSPAN, M., and BAIN, J.: Ultrasonographic examination of scrotal masses. J. Urol. *117:* 185–188 (1977).
6. ROBINSON, D.; ROCK, J., and MENKIN, M.F.: Control of human spermatogenesis by induced changes of intrascrotal temperature. J. Am. med. Ass. *204:* 290–298 (1968).
7. WATANABE, A.: The effect of heat on the human spermatogenesis. J. med. Sci. *10:* 101–106 (1959).
8. ZISKIN, M.C.: Basic principles of diagnostic ultrasound; in GOLDBERG Diagnostic uses of ultrasound, pp. 1–30 (Grune & Stratton, New York 1975).

M. MISKIN, MD, FRCP (C), Departments of Radiology, Medicine and Obstetrics and Gynecology, University of Toronto and Mount Sinai Hospital, *Toronto, Ont.* (Canada)

Pharmacological Therapy in Male Infertility

B. Norman Barwin, D. McKay, E.E. Jolly and R.W. Hudson

Fertility Centre, University of Ottawa, Ottawa General Hospital, Ottawa

Introduction

Involuntary infertility accounts for 10% of all married couples. In up to one-half of these couples abnormalities can be found in the male partner. The study of the problem of male infertility is a relatively young science and at the present time, the results of treating this group of patients is for the most part disappointing. It is important to appreciate that the problem is of concern to both partners and therefore advice and counselling should include both partners.

Since some forms of therapy are known to be effective it is essential that all men suspected of being infertile be properly assessed in order to distinguish those men whose infertility cannot be treated, e.g. Klinefelter's syndrome.

Assessment of the Patient

Clinical

Those aspects of the history and physical examination specifically relevant to the problem are set out in table I. Although in normal subjects, conception has almost always taken place within 6 months of trying for children, it is usual to allow up to a year of infertility before undertaking a full investigation. It is most important to inquire about the frequency and timing of intercourse, as these factors can occasionally provide a simple explanation for the infertility. The frequency of erections and shaving gives some indication about the patient's endocrine status, as does the age of

Table I. Clinical aspects of male infertility

History
Duration of marriage; duration of infertility; contraception
Frequency of intercourse; time of intercourse in relation to wife's menses
Frequency of erections
Frequency of shaving
Medicaments; exposure to heat, toxins, etc.

Past history
Age of onset of puberty
Mumps
Venereal disease
Other endocrine disease
Other general disease, TB, Anemia

Examination
Body habitus, height, arm span
Hair thickness and distribution on scalp, face, pubis, axillae, body and limbs
Presence of gynecomastia
Voice
Size of penis
Consistency, position and size of testes
Spermatic cords
Lying and standing blood pressures

puberty. Since some drugs can cause impotence or inhibit spermatogenesis, the doctor should obtain an account of any treatment the patient is receiving. Similarly, as some illnesses can cause infertility, a past history of these should be sought.

Although not specifically part of the study of the infertile male, the reproductive health of the wife should be assessed, as male and female infertility can coexist.

Examination of the patient may reveal abnormalities such as undescended or small soft testicles. During the examination of the scrotum the spermatic cord, particularly the left side, should be palpated for evidence of varicocele with the patient standing. Small genitalia, a high pitched voice, long limbs, and lack of 'sexual' hair all provide evidence of eunuchoidism. The blood pressure should be taken with the subject both lying and standing to detect postural hypotension, and hence an autonomic neuropathy. The presence of gynecomastia may suggest a diagnosis such as Klinefelter's syndrome. The optic fundi and visual fields should be examined and height and arm span determined.

Table II. Laboratory investigations in infertility

Semen analysis	Measurement of plasma testoterone
Testicular biopsy	and gonadotrophins
Sex chromatin	Culture of urine and prostatic fluid
Chromosome analysis	Serological tests for venereal disease
Skull X-ray film; radiological bone age	Vasography
Tests of hypothalamic pituitary function	

Laboratory Investigations

The more important investigations are set out in table II. Semen analysis which should be performed after 3 days of sexual abstinence at least twice consists of assessment of volume, count of spermatozoa per milliliter, percent motility at 1- and 2-hour intervals from collection and observation of morphology. The presence of pus cells in the seminal plasma may indicate underlying seminal vesiculitis or prostatitis which may be responsible either for decreased sperm count, diminished motility or both.

Random levels of LH and FSH are usually not helpful. High levels of FSH and LH levels are suggestive of a primary testicular disorder. Raised FSH levels with normal concentrations of LH may indicate early spermatogenic arrest. If the plasma testosterone levels are low this may indicate Leydig cell insufficiency. Thyroid function tests should also be ordered.

If hypothalamic or pituitary disease is suspected, an X-ray or tomograms of the pituitary fossa for evidence of pituitary tumor should be performed.

Since infertility may be due to abnormalities of sex differentiation, examination of the sex chromatin of the buccal squames and the chromosome pattern of white cells or fibroblasts may be diagnostically useful in selected cases. Testicular biopsy has not proven to be helpful in the routine management of infertile males with oligospermia or defective sperm. It is indicated in azoospermia to determined whether spermatogenesis is taking place normally hence indicating a possible obstructive lesion. Vasography may then be performed to localize the site of obstruction.

Therapy of Male Infertility

If a specific disease entity has been defined, therapy should be directed to this entity. Investigations fail to define clearly any specific causal factor in the majority of patients. It is also important to appreciate that many

Table III. Causes of male infertility

I	Faults in the testes	testicular agenesis
		cryptorchidism
		chromosomal abnormalities, e.g. Klinefelter's syndrome
		Sertoli cell only syndrome
		chronic illness
		infections, e.g. mumps orchitis
		irradiation and heat
		chemotherapy, toxins
		trauma
		immunological factors
		endocrine disorders, eg. hypothalamic, pituitary disease
		varicocele
		idiopathic
		drugs, e.g. testosterone
II	Faults in transport	congenital absence of vas
		vas ligation (surgical)
		gonorrhea, tuberculosis
		idiopathic
III	Faults in the accessory glands	infections of prostate, seminal vesicles
		immunological factors
IV	Faults of technique	psychological factors
		impotence – psychogenic, medical, neurological, drugs
		premature ejaculation
		infrequent coitus
		use of lubricants

therapeutic successes have been claimed because of a failure of adequate basal observation prior to a therapeutic regime being started. Before any treatment is started, it is important to ensure the wife has been adequately investigated and has no obvious impediments to conception (tab. III).

Specific Treatments

Removal of Specific Offending Agents

Adequate history will elicit to any factors which may interfere with spermatogenesis. Drugs such as immunosuppressive agents, antihypertensives and anabolics may adversely affect sperm quality. Exposure to excessively high temperatures (e.g. frequent sauna baths) and increased scrotal temperature may have a deleterious effect on spermatogenesis.

Urogenital Infections

Reduced sperm count, decreased sperm motility and a reduction in the concentration of acid phosphatase or fructose in the ejaculate is associated with reduced fertility. With persistent leukocytes and cellular debris in the ejaculate infection should be suspected. The finding of a soft tender prostate may also indicate underlying infection. The semen and prostatic secretions (by massage) should be cultured. A gram stain or Papanicolaou stain should also be carried out. The importance of t strains of mycoplasma in the genital tract has been implicated as a cause of both male and female infertility [10]. There is questionable evidence whether eradication of this organism has resulted in an improvement of fertility [6]. GNARPE and FRIBER [10] treated 52 infertile couples with desoxytetracycline (100 mg/day) from the 7th to the 16th day of the cycle to eradicate the organism and reported a 29% pregnancy rate. The combination of trimethoprim and sulfamethoxazole has also been recommended to eradicate this organism [12].

Gonadotrophin Therapy

In hypogonadotrophic hypogonadism, due to Kallman's syndrome or a pituitary tumor, the use of FSH and LH [8] has been recommended. Both gonadotrophins must be used for prolonged periods, the FSH effect being achieved with human menopausal urinary extract of gonadotrophins and the LH effect achieved by HCG.

The acquired group of hypogonadotrophic hypogonadism whose infertility follows pituitary tumors or pituitary surgery may have restoration of spermatogenesis in 50% of patients by the use of FSH and HCG. Doses of 50–100 IU of FSH and 500–1,000 IU of HCG have resulted in fertility after at least 90 days of therapy [21].

LUNENFIELD and SHALKOVSKY-WEISSENBERG [13] in an extensive review were unable to identify the criteria by which patients should be selected for treatment. SCHWARZSTEIN et al. [23] suggested that the patients best suited for treatment are those whose biopsies showed normal spermatogenesis up to the stage of spermiogenesis.

HCG

MISURALE et al. [15] treated patients with normospermia but asthenospermia and reported a 50% pregnancy rate using HCG. FUTTERWEIT and SOBRERO [7] found similar results. HCG may be useful when the ejaculate volume is low or sperm motility decreased.

One of the most interesting ideas in the treatment of male infertility is the combination of HCG and HMG. Recent work by ROSEMBERG [21] has suggested that HCG alone by virtue of its LH activity seems to have a stimulating effect on the spermatogonial phase giving rise to primary spermatocytes. HMG on the other hand stimulated all germinal phases particularly spermatogonial and finally a combination of HCG and HMG seems to result in complete recovery of spermatogenesis and Leydig cell function. POLISCHUK et al. [18] have speculated that a high LH/FSH ratio is necessary for human spermatogenesis. This last concept is supported by MARTIN [14] who has demonstrated normal spermatogenesis in the hypogonadotrophic eunuchoid males with as little as 7–26 IU of FSH daily when supplemented with HCG.

GnRH Therapy

SCHWARZSTEIN et al. [23] have reported the use of LHRH in oligospermia or azoospermia with encouraging results. The disadvantage of LHRH is the fact that it has a short half-life which results in the necessity for frequent injections. Long-acting analogues are being studied and may prove useful in patients with intact testicular function in whom the defect in gonadotrophin secretion results from an absence or deficiency of hypothalamic GnRH.

Anti-Estrogens
Clomiphene Citrate Therapy

There have been many published reports using clomiphene or its isomer cis-clomiphene in male infertility [16, 19]. The rationale for treatment with clomid is based on its gonodotrophin-releasing effect in men [11].

HELLER et al. [11], in a study of normospermic normal men, found at low dosage of 50 mg daily for at least 2 months sperm concentrations increased as well as levels of testosterone, gonadotrophins and estrogens. At high dosages of 200–400 mg/day a decrease in sperm count to azoospermic level was noted despite high levels of gonadotrophins, testosterone and estrogens. PALTI [16] has reported the use of clomid in oligo- and azoospermic males. There were 69 patients in his study treated under four dosage schedules: 12.5, 25, 50 and 100 mg/day for 20, 30 and 60 days with a 47% increase in sperm count and 45% improvement in sperm motility.

Cis-Clomiphene Therapy

Cis-clomiphene is an isomer of clomiphene. REYES and FAIMAN [19] reported statistically significant increases in overall mean sperm concentra-

Table IV. Therapeutic regimens

Indications	Testosterone	FSH	Medication	Dose
Asthenospermia <60% motility	low	normal	mesterolone	100 mg p.o. daily
Oligoasthenospermia <20×10^6/6 <60% motility	low	normal	mesterolone	100 mg p.o. daily
Oligoasthenospermia <20×10^6/6 motility <60%	low	low	clomid	50 mg p.o. daily
Oligoasthenospermia	low	low	HCG	4,000 units i.m. 3× weekly
Azoospermia	—		HMG	75 units i.m. 3× weekly

tions after 6 months of therapy and as a rebound 3 months posttherapy. They found transient increments in serum LH, testosterone and estradiol concentration and a significant decline in circulating FSH levels coincident with sperm concentration peak. Three pregnancies resulted in their series of 16 patients (tab. IV).

Tamoxifen

Tamoxifen is a pure anti-estrogen free from demonstrable intrinsic estrogenic activity in men. COMHAIRE [5] found a twofold increase in sperm concentration and sperm output per ejaculate and three pregnancies resulted in a series of 15 patients. As tamoxifen increases the release of LHRH resulting in testicular stimulation by endogenous LH and FSH, treatment could be more physiological than with exogenous gonadotrophins [4].

Androgens
Testosterone

HELLER et al. [11] originally described the testosterone rebound phenomenon using testosterone propionate at a rate of 25–50 mg on alternate days for periods up to 70 days. ROWLEY and HELLER [20] have claimed substantial improvements in sperm counts up to 2 years after treatment. GETZOFF [9] and CHARNY [3] failed to substantiate these findings as it is well recognized that long-term use of testosterone will induce azoospermia by virtue of its' FSH and LH suppressing effect. The aim of therapy is to induce azoospermia and hope for a rebound increase of sperm count over pretreatment levels [20].

Mesterolone

Mesterolone, a 17β-hydroxy-1-methyl-5-androstan-3-one, is a relatively new androgenic compound which is distinguished from other steroids by the presence of an alkyl group. It is unique among androgenic compounds by its failure to depress LH production. BARWIN et al. [2] reported results of treating 25 oligo- and azoospermic males and 5 impotent males. Dosage was 100 mg daily orally for 1 year. A 33% pregnancy rate was noted in a group of 60 patients with oligospermia although there was a 40% improvement in count and motility. 60% of the impotent males improved. There was no response in the azoospermic group. It should be noted that the azoospermic group all showed germinal arrest patterns or tubular hyalinization on biopsy. PETRY et al. [17] reported similar results. However, SCHELLEN [22] was unable to support these claims.

Thyroid Preparations

Among the treatments traditionally given to a male with subnormal semen are thyroid preparations. There is no evidence that giving thyroxine or triiodothyronine to euthyroid men will in any way affect sperm production [1]. There is no justification for thyroid preparations unless the clinical and laboratory tests support the diagnosis of hypothyroidism.

Concluding Remarks

The treatment of male infertility is frequently unrewarding. It is important that infertile men be thoroughly investigated and that tactful explanation be given to the couple so as not to compromise the relationship between husband and wife. As more knowledge accrues in the understanding of reproduction so will greater strides be achieved in the treatment of male infertility with subsequent improvement of semen analysis and pregnancy rates – the ultimate aim of therapy.

References

1. ARRATA, W.S.M.; ARRONET, G.J., and DERY, J.P.: The subfertile male. Fert. Steril. 20: 460 (1969).
2. BARWIN, B.N.; CLARKE, S.D., and BIGGART, J.D.: Mesterolone in the treatment of male infertility. Practitioner 211: 669–674 (1973).

3 Charny, C.W.: Treatment of male infertility with large doses of testosterone. J. Am. med. Ass. *160:* 98 (1956).
4 Comhaire, F. and Dhondt, M.: Influence of current modes of treatment in male infertility on the hypothalamo-pituitary testicular function; in Schellen Releasing factors and gonadotropic hormones in male and female sterility, p. 117 (European Press Medicon, Ghent 1975).
5 Comhaire, F.: Treatment of oligospermia with tamoxifen. Int. J. Fert. *21:* 232–238 (1976).
6 De Louvois, J.; Blades, M.; Harrison, R.F.; Hurley, R., and Stanley, V.C.: Frequency of mycoplasma in fertile and infertile couples. Lancet *i:* 1073 (1974).
7 Futterweit, N. and Sobrero, A.J.: Treatment of normogonadotropic oligospermia with large doses of chorionic gonadotropin. Fert. Steril. *19:* 971 (1968).
8 Gemzell, C. and Kjessler, B.: Treatment of infertility after partial hypophysectomy with human pituitary gonadotrophins. Lancet *i:* 644 (1964).
9 Getzoff, P.L.: Clinical evaluation of testicular biopsy and the rebound phenomenon. Fert. Steril. *6:* 465 (1955).
10 Gnarpe, H. and Friberg, J.: Mycoplasma and human reproductive failure. Am. J. Obstet. Gynec. *114:* 727 (1972).
11 Heller, C.G.; Nelson, W.O.; Hill, I.B.; Henderson, E.; Maddock, W.O.; Jungck, E.C.; Paulsen, C.A., and Mortimore, G.E.: Improvement in spermatogenesis following depression of the human testis with testosterone. Fert. Steril. *1:* 415 (1950).
12 Keogh, E.J.; Burger, H.F.; Kretser, D.M. De, and Hudson, B.: Non-surgical management of male infertility; in Hafez Human semen and fertility regulation in men, p. 452, chapter 45 (Mosby, St. Louis 1976).
13 Lunenfield, B. and Shalkovsky-Weissenberg, R.: Assessment of gonadotrophin therapy in male infertility. Adv. exp. Med. Biol. *10:* 613 (1970).
14 Martin, F.R.: The stimulation and prolonger maintenance of spermatogenesis by human pituitary gonadotrophins in a patient with hypogonadotrophic hypogonadism. *38:* 431 (1967).
15 Misurale, F.; Cagnazzo, G., and Storace, A: Asthenospermia and its treatment with HCG. Fert. Steril. *20:* 650 (1969).
16 Palti, Z.: Clomiphene therapy in defective spermatogenesis. Fert. Steril. *21:* 838 (1970).
17 Petry, R.; Rausch-Stroomann, J.G.; Hienz, H.A.; Senge, T., and Mauss, J.: Androgen treatment without inhibiting effect on hypophysis and male gonads. Acta endocr., Copenh. *59:* 497 (1968).
18 Polischuk, W.Z.; Palti, Z., and Laufer, A.: Treatment of defective spermatogenesis with human gonadotrophins or testicular development. Fert. Steril. *18:* 127 (1967).
19 Reyes, F.I. and Faiman, C.: Long-term therapy with low dose cis-clomiphene in male infertility: effects on semen, serum FSH, LH, testosterone and estradiol and carbohydrate intolerance. Int. J. Fert. *19:* 49 (1974).
20 Rowley, M.J. and Heller, C.G.: The testosterone rebound phenomenon in the treatment of male infertility. Fert. Steril. *23:* 498 (1972).
21 Rosemberg, E.E.: Gonadotropin therapy of male infertility; in Hafez Human semen and fertility regulation in men, p. 364, chapter 46 (Mosby, St. Louis 1976).

22 Schellen, T.C.M.: Results with mesterolone in the treatment of disturbances of spermatogenesis. Andrologia *2:* 1 (1970).
23 Schwarzstein, L.L.K.; Aparicio, N.J.; Turner, D.; Calamera, J.C.; Mancini, R., and Schally, A.V.: Use of synthetic luteinizing hormone-releasing hormone in treatment of oligospermic men. A preliminary report. Fert. Steril. *26:* 331 (1975).

Dr. B. Norman Barwin, Fertility Centre, University of Ottawa, Ottawa General Hospital, *Ottawa, Ont.* (Canada)

Artificial Insemination and Semen Preservation

B. NORMAN BARWIN

Fertility Centre, University of Ottawa, Ottawa General Hospital, Ottawa, Ont.

Introduction

One of the main problems confronting the gynecologist in the investigation of infertility is the consistent finding of immotile spermatozoa, spermatozoa of low motility and low sperm count in the cervical mucus or semen.

Artificial insemination of the wife with the husband's semen (AIH) and artificial insemination with donor semen (AID) may be indicated in a small proportion of cases and if these are selected with utmost care and the treatment skilfully carried out, it holds a good chance of success. For this reason its' use should be fully understood by all those who seek to help the infertile couple.

History

HUNTER reported successful vaginal artificial insemination in a case of hypospadias at the end of the 18th century [40]. SIMS in 1886 carried out intracervical artificial insemination in 6 women with negative postcoital tests in the cervical mucus; he used the husband's semen obtained from the vagina after intercourse. One patient became pregnant, but aborted. In 1884 in Philadelphia, PANCOAST used AID successfully in one case in which the husband was azoospermic [26]. In the United States in 1890, DICKINSON began using AID in great secrecy [29]. In Great Britain 31 cases of AIH and 15 cases of AID were reported [6]. Since then the use of both AIH and AID has increased as interest in infertility has grown and more is being learned about reproductive physiology.

Incidence

The incidence of artificial insemination is not known but it is probable that it is carried out in many centers. In 1960 the Feversham committee [21] estimated that fewer than 20 practitioners were performing AID in Great Britain and that the number of AID children at that time was not more than 1,150. In the same year it was estimated that AID resulted in between 5,000 and 7,000 births annually in the United States [27].

Ethical and Legal Aspects

Although many people find AIH distasteful, its practice does not appear to present the possibility of any legal complications.

The legal aspect of AID is still under considerable debate. It is doubtful whether a marriage could subsequently be annulled if AID had been carried out with the knowledge and consent of both partners. It is also doubtful whether it could be considered as amounting to adultery. There is no doubt that a child born by AID is illegitimate in law, but the presumption that a child born to a wife during wedlock is legitimate is difficult to rebut, unless it can be shown that the husband is totally infertile or that the blood group of the child is inconsistent with the husband being the father. A child known to be illegitimate must be registered as such at birth; in the case of AID the presumption of legitimacy would enable the person registering the birth to register the husband as father, unless he believed that there was no real possibility that this might be so. One who made a false registration knowingly would be guilty of perjury [30].

Religious Aspects

Most Christian religious authorities are against the production of semen by masturbation but even some Roman Catholics do not object to 'assisted insemination', that is, insemination of the cervix following normal intercourse, using the semen deposited in the vagina. AIH is permissible for Orthodox Jews after 10 years of barren marriage if its therapeutic necessity is attested by two physicians and two Rabbis [27].

For Roman Catholics AID was rejected absolutely by Pope Pius XII in 1949. In a commission set up by the Archbishop of Canterbury in 1948 to discuss this problem it was declared to be contrary to Christian principles and morals. Other Protestant religious bodies have been noncommittal. Orthodox and Reform Jewish Rabbinic authorities who have expressed opinions are not unanimous.

Psychological Aspects

AIH has been more readily accepted by most couples since there exists the possibility of conception occurring and continuance of the genetic and hereditary factors because of insemination with the husband's spermatozoa.

Most couples coming for AID will have experienced considerable psychological trauma when they have learnt that the husband is infertile. The effect of this on each individual and on the marital relationship will depend to a large extent on the emotional maturity of each partner and the stability of the marriage. AID is an ethical medical procedure, but the decision that it is suitable for a given couple should be arrived at by the psychiatrist and gynecologist working together [44]. In a series of 300 AID babies it was found that the recipients of donor insemination were among the happiest of parents [29].

Indications for AIH (table I)

The first step to be taken in the investigation of an infertile couple is to take a full case history, with special inquiry into the marital relationship. Some couples will be aware that intercourse is not taking place properly, either because of impotence or premature ejaculation, with failure of adequate penetration or because of vaginismus or rigid hymen. Occasionally normal intercourse may occur, but with failure of ejaculation or even more rarely, retrograde ejaculation into the bladder because of incompetence of the internal urethral sphincter as a result of injury or neurological disorder. Hypospadias, even in a mild degree, may prevent the semen from being ejaculated high into the vagina even though full penetration occurs.

In all cases the wives should be subjected to a full clinical investigation. Ovulation should be proven on the basis of biphasic basal body temperature records, vaginal cytology, cervical mucus studies, endometrial biopsies and luteal phase serum progesterone levels. Tubal patency should be confirmed by hysterosalpingogram or laparoscopy hydrotubation.

Table I. Indications for AIH and AID

AIH	AID
Female factors	*Male factors*
Vaginismus	azoospermia
Cervical stenosis	oligospermia
Long cervix	testicular atrophy
Retroflexion of uterus	cryptorchidism
Acute anteflexion	orchitis
Extensive cautery	vasectomy
Hypermucorrhea	orchidectomy
Cervical mucus antibodies	irratiation
	psychosexual (impotence)
Male factors	immune reaction
Impotence	genetic factors
Paraplegia	rhesus isoimmunization
Hypospadias	
Prevasectomy storage	
Oligospermia	
Low semen volumen (< 1 ml)	
High semen volumen (> 7 ml)	
Poor liquefaction	

Homologous Insemination

In cases of impotence, premature ejaculation or severe vaginismus treatment should first be directed towards overcoming the difficulty. In cases which do not have a deep psychological cause the practitioner may, by explanation and discussion, help the couple to improve their sexual functioning. In some cases the husband's potency may be improved by small doses of methyltestosterone given during the first half of his wife's menstrual cycle, leading up to her ovulatory time; but more often short-term behavioural therapy or psychotherapy may be necessary. Every effort should be made to overcome sexual difficulties before artificial insemination is considered.

Male indications for homologous insemination are: small volume of the ejaculate (especially when the volume is 1 ml); low sperm density (oligozoospermia) or large volume of ejaculate (more than 7 ml) with low sperm density, two situations in which the utilization of the better portion of a split ejaculate may solve the problem; semen of high viscosity or that does

Table II. Results of AIH when semen analysis showed a 'normal'[1] sperm count

Specialist referral	Indication	n	Inseminations/ Cycle	Result	Success, %
Urologist	retrograde ejaculation	1	4.0	0	0
Neurologist	impotence (paraplegics)	5	3.0	3 pregnant[2]	60
Radiotherapist	seminoma	3	3.5	2 pregnant	66
Psychiatrist	vaginismus (1) impotence (2)	3	2.5	2 pregnant	66
Infertility clinic	cervical mucus factor	18	3.5	13 pregnant[3]	70
Total		30	3.3	20 pregnant	66

[1] $>20 \times 10^6$ spermatozoa/ml and 50% motility.
[2] There was one abortion.
[3] There were two abortions.

not liquefy satisfactorily; anatomic defects, such as hypospadias or epispadias, marked obesity, and chordee or Peyronie's disease, which make penetration impossible and/or deposition of semen inadequate; paraplegia; retrograde ejaculation, whether neurogenic, congenital, or post-operative; and late ejaculation (table II).

In the cases of excessive semen viscosity or semen that fails to liquefy, one could theoretically expect that viscosity could be diminished by the administration of estrogens over a short period of time, not affecting spermatogenesis but interfering with the acessory glands, prostate, and seminal vesicles and modifying and diminishing their secretion and possibly semen viscosity. The use of tyloxapol (Alevaire), a mucolytic solution, in a 1:1 (semen:tyloxpanol) proportion, or the forcible passage of the ejaculate through an 18- or 19-gauge hypodermic needle four or five times may diminish the viscosity [2].

If the postcoital test is found to be negative on several occasions, this small volume of semen is injected directly into the cervix. AIH has been tried when the husband is oligospermic or when the wife's cervical mucus is impermeable to spermatozoa despite therapy. A 55% success rate using intrauterine AIH has been reported [7].

Table III. Results of AIH when semen analysis showed oligospermia

Specialist referral	Indication	n	Inseminations/Cycle	Results	Success, %
Infertility clinic	low grade oligospermia ($<10 \times 10^6$ sperm/ml)	8	6	3 pregnant[1]	38
Infertility clinic	High grade oligospermia ($<20 \times 10^6$ sperm/ml)	12	3	6 pregnant[1]	50
Total		20	4.5	9 pregnant	45

[1] There were two abortions.

Intrauterine insemination may overcome any cervical factors such as spermagglutinins or antisperm antibodies present in the cervical mucus [32]. A 32% pregnancy rate using AIH because of persistent or unexplained poor postcoital tests despite a normal semen picture has been reported by STEIMAN and TAYMOR [42].

The use of split ejaculates to bring about a natural concentration of sperm does provide a rationale for AIH in oligospermia.

FARRIS and MURPHY [19], by intrauterine insemination of the best portion of the split ejaculate, had 13 pregnancies in 100 couples. AMELAR and HOTCHKISS [2] had 13 pregnancies in 23 patients (56%), using split ejaculates for sperm counts of less than 40 million/ml, but no mention was made of the insemination technique. Using split ejaculates injected either into the cervix or uterus, PEREZ-PELAEZ and COHEN [31] had 10 pregnancies in 38 couples (26.3%).

BARWIN [7] had 20 cases of infertility in which the sperm count was less than 20 million/ml. The semen was stored in liquid nitrogen, and some specimens were pooled to improve the count. All inseminations were intrauterine, and 45% pregnancies were achieved (table III).

Donor Artificial Insemination

Before beginning donor insemination, both husband and wife should be interviewed and an assessment made of their personalities, their reaction to each other, and the suitability of this treatment for them. The legal and ethical sides of the matter should be carefully explained to them. They should both sign a consent form in the presence of the physician. The practical aspects of the treatment should be explained in detail, along with the possibilities of success.

Normal fertility of the wife in cases of infertility due to male factors cannot be assumed; as many as 75% of women referred for donor insemination have been found to have some stigmata of infertility [5]. If conception is to occur, concurrent therapy in the wife is essential. In the absence of any obvious cause of infertility, AID can be tried for 2 months before fertility tests are begun; a 40% conception rate during this time can be expected [27].

Selection of Donor

The donor and recipient should always be completely anonymous. He should be of the same race as the recipient, have the same social and intellectual background, and have a negative drug and venereal disease history. The donor's blood group should preferably be the same as that of the husband or the wife. If the wife is Rh negative, an Rh negative donor should be used. There should be a 48-hour period of abstinence prior to donation.

In the United States, young hospital physicians, preferably with children, or medical students are used as donors. In Great Britain, semen donation is often voluntary, and donors may sometimes be found among the husbands of patients who have been successfully treated for infertility resulting from other factors. Tests should be made for syphilis, *Neisseria gonorrhoeae*, *Haemophilus vaginalis*, candida, trichomonas, and hepatitis virus type B.

Indications for AID (table I)

AID is most often considered in a marriage in which the wife appears to be fertile and the husband is sterile or potentially so, based on spermatozoa count, motility, or morphology. In these cases, it may be the only alternative to adoption. AID may also be helpful in Rh incompatibility problems in which the husband is homozygous Rh positive and the wife has given birth to one or more hydropic fetuses. It is useful in cases in which the husband is known to carry the genes of some hereditary disease that would make fatherhood inadvisable.

The diagnosis of male infertility should be reached only after repeated semen tests and consultation with a specialist who is experienced in this area and who will be able to discuss with the husband the possibility of his condition being improved by treatment.

The use of donor insemination has generally resulted in a conception rate between 55 and 80%. Almost 90% of the women who eventually become pregnant do so within 6 months. A large number of pregnancies occur during the first month of insemination (37%), and 75% of the patients from the same series became pregnant during the first 3 months of treatment [27].

Timing of Insemination

Before initiating artificial insemination, the average midcycle day is calculated from the observation of cycles for 3 consecutive months. The patient is instructed to contact the physician 2 days before the presumed midcycle day, i.e. 2 days before the rise of basal body temperature (BBT). Successful insemination usually occurs 12–36 h before the rise of BBT [29]. If the temperature has not risen 2 days later and if the mucus still appears preovulatory, insemination is repeated. Three inseminations per month (days 11, 13, and 15 in a 28-day cycle and days 13, 15, and 17 in a 30-day cycle) have been recommended [27].

Accurate estimation of ovulation is of paramount importance in achieving conception with artificial insemination. It is now accepted that the optimal time for insemination is the day of ovulation or 1–2 days before ovulation [11].

In BARWIN's [7] study, inseminations began 2–3 days before the expected temperature rise on the basal body temperature chart and continued daily until the temperature rose to its postovulatory level. This resulted in an increase in the number of inseminations per cycle but perhaps accounted for the reduction in the number of insemination cycles to achieve a pregnancy [7].

The intravenous injection of conjugated estrogen (Premarin) greatly increases the success rate of AIH [22]. It induces luteinizing hormone (LH) secretion by increasing the estrogen 'surge'. The number of inseminations per cycle is reduced, and accuracy in predicting ovulation is increased, as indicated by the pregnancy rate [7] (table IV).

Table IV. Number of cycles to conception following AIH

	Months								
	1	2	3	4	5	6	7	8	9
Number pregnant	6	7	9	5	2	1	–	–	1
Number of inseminations/cycle	2	3	4.5	2.5	3.0	3.0	3.0	3.0	2.0
Proportion, %	19	23	28	18	6	3	0	0	1

Intramuscular injection of human chorionic gonadotropin (hCG) (5,000 U) in the late follicular phase of the artificial insemination cycle has been used successfully [24]. Cervical mucus ferning and spinnbarkeit vaginal cytology together with the BBT chart record are valuable adjuncts in timing ovulation [14, 17].

Technique

Intrauterine insemination is used when the endocervical mucus is impermeable to spermatozoa or in cases of oligozoospermia [7, 27, 29]. A sterile mixing needle is introduced into the uterine cavity, and 0.1–0.3 ml of semen and glycerol-egg yolk-citrate buffer is injected. In the presence of a cervical mucus factor, the cervical mucus is aspirated with the subsequent instillation of donor or homologous semen in the uterine cavity. It seems logical to suppose that semen with relatively low spermatozoa counts inseminated into the uterine cavity would result in numbers of spermatozoa equivalent to those reaching the uterine cavity after normal coitus, provided motility is good [20]. Objections have been raised to intrauterine insemination because of the complication of violent cramps, as well as the risk of intrauterine infection and chemical salpingitis [16, 43]. For this reason, it is important that the pH of the buffer be 7.2–7.4 and the volume be small (0.2–0.3 ml).

Intracervical insemination is used in cases of cervical stenosis when the external os presents a barrier to spermatozoa migration. The cannula is inserted to the midposition of the cervical canal and the semen injected slowly, a few drops at a time.

In intravaginal insemination the whole of the specimen is injected into the vaginal vault without the necessity of exposing the cervix; this may be carried out by the couple themselves in some cases of coital difficulty.

The cervical cup method involves use of a plastic cup that is placed over the cervix; a whole or split ejaculate is injected into the cup by means of a cannula. The patient may rise from the couch immediately afterward, as there will be no efflux of semen. The cup is left in position for at least 8 h and is removed by the woman herself [38].

Cervicovaginal, or intracervical, insemination is the method most often used. The patient is placed on the examination couch with her hips slightly raised, and a clean, dry, bivalve speculum is inserted into the vagina, and the cervix is exposed. The semen is drawn up into a sterile syringe or a cannula attached to a 5-ml syringe; 1 drop is injected into the mucus just within the cervical canal and the remainder instilled onto the cervix and vaginal vault. The speculum is removed, and the patient remains recumbent for about 30 min.

Semen Preservation

There has been tremendous interest in the use of frozen semen in human insemination since it was discovered that glycerol in the suspending medium can act as a cryoprotective agent [34]. Since 1953, when BUNGE and SHERMAN [15] found that glycerol-treated human spermatozoa after freezing and thawing could result in fertilization, great interest has been focused on sperm freezing and the establishment of sperm banks in human medicine. The freezing of human sperm has not been altogether successful, since the pregnancy rate after inseminations with frozen sperm has been lower than the rate following inseminations with fresh sperm [39]. The problem in sperm preservation is the low recovery index (the depression of postfreezing motility in relation to fresh sperm motility). Crystallization and respiratory shock occurring after freezing cause some 25% loss in fertilizing power.

The great advantage of frozen semen is the availability of the specimens for inseminations with a ready source of potentially fertile semen from donors whose physical characteristics, genetic history and blood group are known. The practicability of having banked semen available for multiple inseminations with ovulation induction, preservation of semen prior to vasectomy, testicular surgery or radiation and the concentration and preservation of the spermatozoa of the oligozoospermic husband is of obvious practical importance.

Cryopreservative Agents

The common approach for semen preservation is to dilute whole semen with either glycerol or a semen extender (glycerol-egg yolk-citrate), to prevent cryoinjury, and then to freeze it, either rapidly within a very few minutes or slowly over approximately 30 min. The frozen sample is then submerged in liquid nitrogen for storage. The great majority of the results have indicated that approximately 30–70% of spermatozoa that were motile prior to freezing regained their motility after thawing.

Various concentrations of glycerol (5–10%) have been mixed with semen prior to freezing [37]. Other investigations have added other constituents with the glycerol with the rationale that the spermatozoa would be in a more physiological medium better able to meet metabolic needs.

For 1,000 ml of buffer, the following concentration were required: germ-free egg yolk, 200 ml; glycerol, 140 ml; glucose (5%), 264 ml; sodium citrate (2.9%), 396 ml; glycol, 20 g [12]. The semen to be stored is then mixed in a 1:1 ratio with the protective medium [7]. FRIBERG and GEMZELL [23] reported 10 pregnancies after inseminations with glycerol-egg yolk-citrate treated semen.

In order to improve sperm motility of frozen semen 7.2 mM of buffered caffeine have been added prior to semen preservation with 40–80% increase of sperm motility in semen specimens of low initial motility (30–60%) [4]. The use of rapid-rate freezing to produce frozen sperm pellets and the addition of one frozen pellet of buffered caffeine (7.2 mM) to five sperm pellets before thawing is recommended. For slow-rate freezing in pellets, the addition of a caffeine solution (final concentration, 7.2 mM) to the sperm and protective medium should be taken into consideration.

Improved motility of human spermatozoa after freeze preservation has also been reported [25]. The sperm freeze preservation procedure involved dilution of the semen samples and separated sperm fractions with a cryoprotective semen extender, and freezing and thawing in a conventional manner. Postthaw motility and percentage survival was substantially higher for the separated fractions than for the parent semen.

Apart from the ability to produce highly motile spermatozoa the efficacy of the sperm separation technique in isolating the Y chromosome is unresolved [18, 35].

Freezing Apparatus and Process

Several methods have been used for freezing semen, the basic difference among these methods is the rate of cooling of the semen which may be either rapid or slow. All methods ultimately utilize liquid nitrogen (–196°C) for the storage process [3, 13, 33, 39]. The semen is given time to liquefy and the buffer added in equal volumes with the semen. This procedure is carried out under sterile conditions. The diluted semen is cooled to 18°C before freezing. Plastic tubes known as 'straws' are filled with semen, each straw being 10 cm long and having a volume of 0.5 ml. The patient's name is stamped on the straws and each straw heat-sealed. The straws are then placed in a metal rack and frozen in the vapor of a large liquid nitrogen vessel to –196°C. On average, 15 straws are required for each diluted specimen. A similar method has been described by BEHRMAN and SAWADA [13]. The nitrogen vapor technique is the most commonly used freezing process; this is an example of a rapid freeze [23, 33]. Aliquots of semen-medium mixture are placed in plastic straws, glass ampoules, or plastic vials. The mixture is then suspended over liquid nitrogen in a closed system for 15–30 min, with the container then being plunged beneath the liquid nitrogen.

The biological freeze technique (a programmed slow freezing process) has been developed [12]. The cooling schedule is as follows: 1°C/min from room temperature to 2°C, 5–7°C/min between 2 and 20°C, so the semen passes through the heat of fusion within 3 min, and then approximately 10°C/min down to –80°C. At this point the mixture is placed beneath liquid nitrogen for storage. The egg yolk-citrate-glycerol medium is used as the cryoprotective agent with this method of freezing. The originators believe that there is a better return of motility when thawed. There is renewed interest in the use of pellets for storage of semen [3]. However, semen in 'straws' has been found to freeze more evenly, presumably because the temperature changes are evenly spread throughout the diluted semen and better recovery rates of motile spermatozoa have been recorded. Another advantage of the straw technique is its ability to withstand the strain involved in freezing at very low temperatures [7].

Post-Thaw Recovery

The speed with which the mixture passes through these various ranges is critical for the ultimate postthaw motility recovery [36]. Any method employing liquid nitrogen assumes a rapid transition of the semen-medium mixture through the range of recrystallization, the rate achieved being between 16 and 25°C/min. The respiration of spermatozoa following thawing has indicated that the rate of cooling through the various temperature renges may cause less damage to the spermatozoa when the mixture is taken through the range of temperature shock ta a slow rate (1°C/min) than when it is rapidly cooled (15–20°C/min) [37].

The freezing of semen always involves some loss of motility [8, 9]. Loss of motility in the range of 25–45% is commonly reported with the freeze-thawing of human semen [9]. There is no method of predicting how a good semen specimen will respond to cryoinjury [10].

Length of Storage

A progressive, significant loss of motility has been reported in samples that had been stored for longer than 36 months [41]. The longest reported experience with frozen stored human semen found no decrease in postthaw motility with time, and found there was a 70% survival of spermatozoa motility after 1, 4, and 10 years [39].

Chromosomal and Morphological Changes

There appears to be no chromosomal damage done to spermatozoa by the freezing process as used in the cattle industry where inseminations with frozen semen have been performed over many years. Morphological changes seen with the electron microscope following freezing are hard to interpret because of similar alterations occasionally seen in spermatozoa in the fresh state, although, damage has been noted in the middle piece, with swelling of the mitochondria and wrinkling and rupture of the acrosome. The incidence of abnormalities among persons using artificial insemination is no higher than that of the general population [28].

Conclusion

Fewer infants are available for adoption because of the increased availability of contraception and the liberalization of abortion laws. The poor response to medical or surgical treatment in male fertility has increased the need for artificial insemination. The proper assessment of the couple, the timing of ovulation and careful selection of donors is of the utmost importance. Although the efficiency of insemination has been greatly aided by progress in semen preservation, the conception rate using fresh semen is higher.

References

1　AMELAR, R.D.: Coagulation, liquefaction and viscosity of human semen. J. Urol. *87:* 187 (1962).
2　AMELAR, R.D. and HOTCHKISS, R.S.: The split ejaculate. Its use in the management of male infertility. Fert. Steril. *16:* 46 (1965).
3　BARKAY, J.; AUCKERMAN, H., and HERMAN, M.: A new practical method of freezing and storing human sperm and a preliminary report on its use Fert.. Steril. *25:* 399 (1974).
4　BARKAY, J.; ZUCKERMAN, H.; SKLAN, D., and GORDON, S.: Effect of caffeine on increasing the motility of frozen human sperm. Fert. Steril. *28: 2:* 175 (1977).
5　BARTON, M.: Artificial insemination. Stud. Fert. *7:* 99 (1955).
6　BARTON, M.; WALKER, K., and WEISNER, B.P.: Artificial insemination. Br. med. J. *1:* 40 (1945).
7　BARWIN, B.N.: Intrauterine insemination of husband's semen. J. Reprod. Fert. *36:* 101 (1974).
8　BARWIN, B.N. and BECK, W.W.: Artificial insemination and semen preservation; in HAFEZ Human semen and fertility regulation in men, pp. 429 (Mosby, St. Louis 1976).
9　BECK, W.W.: Artificial insemination and semen preservation. Clin. Obstet. Gynec. *17:* 115 (1974).
10　BECK, W.W. and SILVERSTEIN, I.: Variable motility recovery of spermatozoa following freeze preservation. Fert. Steril. *26:* 863 (1975).
11　BEHRMAN, S.J.: Artificial insemination. Fert. Steril. *10:* 248 (1959).
12　BEHRMAN, S.J. and ACKERMAN, D.R.: Freeze preservation of human sperm. Am. J. Obstet. Gynec. *103:* 654 (1969).
13　BEHRMAN, S.J. and SAWADA, Y.: Heterologous and homologous inseminations with human semen frozen and stored in a liquid-nitrogen refrigerator. Fert. Steril. *17:* 457 (1966).
14　BILLINGS, E.L.; BILLINGS, J.J.; BROWN, J.B., and BURGER, H.G.: Symptoms and hormonal changes accompanying ovulation. Lancet *1:* 282 (1972).
15　BUNGE, R.G. and SHERMAN, J.K.: Fertilizing capacity of frozen human spermatozoa. Nature, Lond. *172:* 767 (1953).
16　CARRUTHERS, G.B.: Husband and donor insemination in infertility. Med. J. Gynecol. Sociol. *5:* 13 (1970).
17　COHEN, M.R.; STEIN, I.F., and DAYE, B.M.: Optimal time for therapeutic insemination. Fert. Steril. *7:* 141 (1956).

18 Ericsson, R.J.; Langevin, C.N., and Nishino, M.: Isolation of fractions rich in human Y sperm. Nature Lond. *246:* 421 (1973).
19 Farris, E.J. and Murphy, D.P.: The characteristics of the two parts of the partitioned ejaculate and the advantages of its use for intrauterine insemination: a study of 100 ejaculates. Fert. Steril. *11:* 465 (1960).
20 Faundes, A.; Croxatto, H.; Mendel, M., and Vera, C.: Sperm migration in the female genital tract. Excerpta Med. Int. Congr. Ser. *234:* 48 (1971).
21 Feversham Committee: Report on human artificial insemination. Br. med. J. *ii:* 379 (1960).
22 Foldes, J.J.: Artificial insemination. Int. J. Fert. *10:* 248 (1972).
23 Friberg, J. and Gemzell, C.: Inseminations of human sperm after freezing in liquid nitrogen vapors with glycerol or glycerol-egg-yolk-citrate as protective media. Am. J. Obstet. Gynec. *116:* 330 (1973).
24 Fuchs, K.; Brandes, J.M., and Paldi, E.: Enhancement of ovulation by chorigon for sucessful artificial insemination. Int. J. Fert. *11:* 211 (1966).
25 Glaub, J.; Mills, R., and Katz, D.F.: Improved motility recovery of human spermatozoa after freeze preservation via a new approach. Fert. Steril. *27:* 1283 (1976).
26 Gregoire, A.T. and Mayer, R.C.: The impregnators. Fert. Steril. *16:* 130 (1965).
27 Guttmacher, A.F.: The role of artificial insemination in the treatment of sterility. Obstet. Gynec. Surv. *15:* 767 (1960).
28 Iizuka, R.; Sawada, Y.; Nishina, N., and Ohi, M.: The physical and mental development of children born following artificial insemination. Int. J. Fert. *13:* 24 (1968).
29 Kleegman, S.J. and Kaufman, S.A.: Infertility in women (Blackwell, Oxford 1966).
30 Law Society: On artificial insemination. Br. med. J. *ii:* 584 (1959).
31 Parez-Pelaez, M. and Cohen, M.R.: The split ejaculate in homologous insemination. Int. J. Fert. *9:* 25 (1965).
32 Parrish, W.E. and Ward, A.: Studies on cervical mucus and serum from infertile women. J. Obstet. Gynaec. Br. Commonw. *75:* 189 (1968).
33 Perloff, W.H.; Steinberger, E., and Sherman, J.K.: Conception with human spermatozoa frozen by nitrogen vapor technic. Fert. Steril. *15:* 501 (1964).
34 Polge, C.; Smith, A.U., and Parkes, A.S.: Revival of spermatozoa after vitrification and dehydration at low temperatures. Nature, Lond. *164:* 666 (1949).
35 Ross, A.; Robinson, J.A., and Evans, H.J.: Failure to confirm separation of X- and Y-bearing sperm using BSA gradients. Nature Lond., *253:* 354 (1975).
36 Sawada, Y. and Ackerman, D.R.: Use of frozen human semen; in Behrman and Kistner Progress in infertility (Little, Brown, Boston 1968).
37 Sawada, Y.; Ackerman, D.R., and Behrman, S.J.: Motility and respiration of human spermatozoa after cooling to various low temperatures. Fert. Steril. *18:* 775 (1967).
38 Semm, K.; Brandl, E., and Mettler, L.: Vacuum insemination cap; in Hafez Human semen and fertility regulation in men, p. 439 (Mosby, St. Louis 1976).
39 Sherman, J.K.: Synopsis of the use of frozen human semen since 1964: state of the art of human semen banking. Fert. Steril. *24:* 397 (1973).
40 Shields, F.E.: Artificial insemination as related to female. Fert. Steril. *1:* 271 (1950).

41 SMITH, K.D. and STEINBERGER, G.: Survival of spermatozoa in a human sperm bank. Am. med. Ass. *223:* 774 (1973).
42 STEIMAN, R.P. and TAYMOR, M.L.: Artificial insemination homologous and its role in the management of infertility. *28:* 146 (1977).
43 WARNER, M.P.: Five hundred cases of human artificial insemination; analysis of 44 years experience. Excerpta Med. Int. Congr. Ser. *234:* 48 (1971).
44 WATTERS, W.W. and SOUSA-POSA, J.: Psychiatric aspects of artificial insemination (donor). Can. med. Ass. J. *95:* 106 (1966).

B. NORMAN BARWIN, Fertility Centre, University of Ottawa, Ottawa General Hospital, *Ottawa, Ont.* (Canada)

Psychiatric and Psychological Aspects of Male Infertility

David M. Berger

Mount Sinai Hospital, Reproductive Biology Unit, University of Toronto, Toronto, Ont.

Introduction

The purpose of this chapter is to discuss the psychiatric and psychological aspects of male infertility from the perspective of a psychiatrist consulting to a reproductive biology clinic. Within this context the psychological aspects of male infertility are closely intermeshed with issues pertaining to the couple as a unit and issues pertaining to the clinical setting in which such problems are encountered.

The Infertility Work-Up as Trauma

Whether conscious or not, anxiety, frustration, and fears of inadequacy are likely present in every couple seeking help for an infertility problem. The work-up in which they are about to engage will, in itself, have elements that are conflictful. It will probe into the couple's sex life, a highly private area. It will demand performance; the man especially will be required to produce semen specimens. The latter has been reported to precipitate impotence, a circumstance that requires immediate treatment [2]. The work-up may end with the couple discovering that they cannot have children, and with one of them named as the individual 'at fault'. No wonder that patients drop out or miss appointments or even become pregnant. In our experience 20% of successful pregnancies occur prior to completion of the work-up. It is helpful to discuss at the start with the couple, what they can expect to happen during the work-up, to share, whether or not they appear concerned, the staff's awareness that the work-up may engender anxiety, and to explore the motivations and mixed feelings of *each* partner. A couple may admit to ambivalence about having children. In the man it often takes the form of financial concerns. How realistic this is has to be considered.

The marital interaction has to be examined. Although both partners may have mixed feelings about having children, this conflict often comes to the surface with one partner expressing a desire to have children and the other against it. In one couple, for example, the husband seemed reluctant to have children but felt he ought to comply with his wife's wishes.

Whenever the wife talked about children, she became somewhat sad and wondered whether her husband would be an adequate father. On further discussion, the wife's fear turned out to be a screen for her own concerns that she would be an inadequate mother. In order to help his wife keep this knowledge from herself (i.e., to help her deny her ambivalence), and vaguely sensing the concern and depression that having a child might arouse in her, a depression he might not be able to cope with, the husband chose a negative stance on the issue of having children.

The psychiatrist's role is not to make decisions for the couple but by relating present concerns to past events and by examining the marital interaction, to weed out distortions and issues that are unreal so that the couple can come to a decision realistically. A couple's deciding against having children, it should be noted, is just as valid a resolution as a decision in the opposite direction. That we are all wholeheartedly in favor of having children is a social myth to which the physician hopefully does not adhere. One of the dangers of a reproductive biology clinic is that, in its manner of functioning, it may place an implicit expectation on the client: to become pregnant.

A more subtle issue is the fact that prior to the work-up the couple may have fashioned a myth reinforced by early unconscious conflicts in *each* partner, to explain the infertility, a myth that the infertility work-up may destroy. On admission for example, one couple admitted to a shared belief that it was the wife's medical problems that were the cause of the infertility. It was based on shaky evidence. The wife was once discovered to have a minor gynecological problem. Further discussion revealed that the husband had a long history of compensatory hypermasculine behavior. The wife came from a home in which mother was dominant, a pattern she vowed never to repeat. The shared belief can be seen here to satisfy the needs of both partners.

If the work-up discovers that the husband is azoospermic, it may be quite unsettling to the relationship. Unconscious fears that a necessary myth may be destroyed can contribute to a couple's reluctance to participate in an infertility work-up.

Psychosexual Problems Masquerading as an Infertility Problem

A variety of psychosexual problems may present as a problem of infertility. This possibility should be considered whenever the desire to have children seems less than urgent. Applying for an infertility work-up can be a masked request for sexual or marital counselling but there may be other unspoken issues. At the initial interview, one couple for example, stated that they did not want children immediately but were anxious to find out whether the husband had an adequate sperm count. The husband's history revealed that he had recently been treated for depression in a psychiatric hospital. It soon became apparent that his concern about his sexual functioning represented a displacement of his need for reassurance that he was 'all right'.

Psychological Factors as the Cause of Infertility

Perhaps the simplest and most understandable manner in which psychological conflicts may cause infertility is through their effect on sexual performance. Sexual problems are often classified according to the particular performance problem. In men these are listed as impotence (i.e., failure to achieve or maintain an erection), ejaculatory problems, ranging from premature ejaculation to none at all, and problems relating to sexual frequency. It has been suggested that as well as low frequency of sexual activity, excessive frequency may also be a cause of infertility [4].

Perhaps a more useful classification is one in which sexual problems are described as primary or secondary (i.e., has the performance problem always been present or is it of recent onset?). Secondary problems have a better prognosis.

Short-term therapeutic techniques include support, encouragement and education. Perhaps the best known brief therapeutic technique is that of MASTERS and JOHNSON [8]. These workers regard sexual functioning as a natural and reflexive physiological phenomenon that can be disrupted by anxiety, depression or stress. Their brief but highly concentrated psychotherapy involves male and female co-therapists working with the couple as a unit and focuses on sexual performance. In addition to specific suggestions and improving communication between partners on sexual issues, they employ a program of systematic desensitization to diminish performance anxiety and alter the partner's role from that of spectator to one of participant.

Particular masturbatory techniques (e.g., the squeeze technique for premature ejaculation) are also employed.

A psychoanalytically oriented therapist would be interested in relating the onset of the symptom to a particular stress, to attempt to understand this stress, with the help of symbolic material, as an event that has caused the arousal of unconscious conflicts originating from the patient's childhood, and also to consider the secondary gain derived from the symptom for the individual and the couple.

All of this implies that a careful and detailed sexual history is essential. Does the husband abstain from intercourse during the wife's most fertile periods? Is the man who complains of impotence able to maintain an erection when he masturbates or reads a Playboy magazine? In one couple the husband had misplaced the thermometer during their first attempt at keeping a temperature chart and had never remembered to replace it. Issues such as these help delineate the problem and clarify whether subtle and unconscious behavior is working against conception.

The role of the psychiatrist is much more difficult when problems of actual sexual performance have not been discovered. Although lacking in empirical evidence, the route often taken is to search for fears, resistances and unconscious conflicts relating to pregnancy and child rearing [1]. The possibilities are too numerous to list. One must wonder whether secondary motives lie behind the wish to have children. Is the much wanted child an attempted cure for a failing marriage, one that both partners are ambivalent about maintaining? Are there anxieties about pregnancy, about the capacity to parent? Although such concerns are often attributed to women, they are not absent in men. Myths such as the story that Athena sprang from the forehead of Zeus and the ancient ritual of couvade suggest that the male has had a long and intimate psychological involvement with childbirth. A recent military study of soldiers whose wives were pregnant revealed a variety of unconscious concerns about the expected child and an increased incidence of psychopathy during the pregnancy [3]. Rivalry with the child for the wife's attention, competitive strivings, fear of harming the child, the arousal of maternal (hence unacceptable, unmanly) feelings are only a few possibilities of resistances in the male. Historical details relating to the individual's own childhood, especially the birth of siblings, are often informative.

If there is evidence of a major psychiatric disorder, the physician is of course obligated to treat it with the hope that in some way the problem of infertility will be corrected. However, in the absence of severe psychopathology, the psychiatrist is confronted with a number of difficult questions.

Such conflicts are present in all of us. How severe must they be before one regards them as pathological? If during the course of the work-up biological abnormalities are discovered, they of course take precedence. But if not, should one attempt to deal with such conflicts? Can one assume that such conflicts can lead to infertility in the absence of a sexual disorder?

Here the psychiatrist is in the position of the lone soldier holding off the enemy until reinforcements arrive. There is evidence that the pituitary-adrenal axis provides a pathway connecting emotions and sexual behavior. Animal studies have shown that sexual behavior is affected by the functional interplay of classical hormones and monoamines [7]. In humans the treatment of parkinsonism with L-dopa may have an aphrodisiac effect [5]. In regard to fertility, the psychiatrist has two conceptual models to choose from. The first is that conflicts affect the interplay of monoamines and classical hormones, and lead to infertility via, for example, seminal changes. A recent report suggests that stress is associated with a decreased sperm count [9]. The second model (which may be preferred by biologically oriented psychiatrists) suggests that both psychological problems and infertility have a common organic basis. 'Macroscopic' studies which correlate personality traits or mood levels with infertility are difficult to interpret (for example, the fact that infertile couples exhibit depression and lowered self esteem) [10]. It is unclear how these parameters interact or whether one is of greater significance etiologically.

Despite these many difficulties, it seems correct and useful to bring to the couple's awareness any conflicts that may be working against their becoming pregnant. When possible or appropriate, pointing out the connection of present fears to past events can help diminish the impact of the fear. In one couple for example, a husband's concerns about having children related to the fact that his father had died shortly after his mother had become pregnant.

Helping the Couple Deal with the Outcome of the Infertility Work-Up

The couple that has discovered that they cannot have children may require help in mourning the loss. Inability to have children is a real loss with mourning and depression as a natural consequence. Empathy is helpful but not sufficient. As in all mourning processes, secret vices, guilt, shames and inadequacies may attach to the mourning process, adding to the burden. It is not sufficient to simply accept the fact that the infertile couple is depressed, 'for good reason' and not press the matter further, especially where the depression is excessive. The physician must say 'Yes, it is natural to be depressed, but let's not stop talking about it'. He has to bear with the patient because private concerns do not come out easily. If he is impatient, he may not discover for example that a married man attributes his azoospermia to the fact that he masturbates.

In our experience men who are discovered to be azoospermic often react to this news with depression and impotence. The associated self-concept is one generally described as a feeling of unmanliness. An important first measure is to reassure the patient that infertility has no biological effect on sexual performance or on any other social parameter associated with masculinity. If symptoms or marital difficulties persist, one can assume that early unconscious conflicts have been aroused and that the symptoms serve the function of secondary gain either intrapsychically or within the marital nexus. The 'guilty' member invariably suffers a blow to his or her self-esteem but the reaction of the 'exonerated' member is equally important. A wife may subtly play up her husband's low sperm count to defend against her own sense of inadequacy. A husband's smothering concern for an anovulatory wife may be a vehicle for resentment or a reaction against the fear that he has somehow damaged her. Similarly one must be watchful for the possibility that a fixed belief shared by the partners and a necessary complement to each partner's unconscious conflicts, may be shattered by the results of the work-up.

However marital disharmony may occur, rage at the partner who initiated the work-up may result. Hence the necessity to engage both partners equally at the start of the work-up.

Often at the completion of a work-up the couple may be confronted with important decisions as to how to proceed in their quest for a child. Should they adopt? Is artificial insemination by a donor the best course to follow? Where the individuals are in conflict even after a clear and frank exposition of the situation by their physician, psychiatric intervention is useful – again, not to offer direction but to help weed out unrealistic concerns and distortions that may have attached to the decision-making process.

A not uncommon outcome of an infertility work-up is that no definite cause for the infertility can be pinpointed. The so-called physically 'normal couple' is an extreme example of this. Psychiatric involvement can serve two useful functions: first, to reassess in fine detail sexual behavior and conflicts about pregnancy that may have been overlooked; and second, to share and discuss the couple's uncertainty and perhaps to help plan a course of action for the future.

Discussion

The multiplicity of clinical issues creates a problem for the psychiatrist. He cannot be in all places at all times. As well, psychiatric treatment is often of longer duration than medical treatment. The following recommendations for psychiatric involvement in a reproductive biology clinic are made in the light of these difficulties.

(1) An *initial* interview to deal with anxiety, to discuss frankly what might be expected during the infertility work-up and to explore sexual and marital problems, concerns and unrealistic beliefs, should involve *both* partners. The partners should be seen together and individually. In our experience individual interviews are most productive when the interviewer is of the same sex as the client. In our clinic the intake function is performed by a nurse and a social worker. Initial assessments are discussed with a psychiatrist at a regular intake conference. Couples that seem to manifest an emotional disorder are then seen together and individually by the psychiatrist with the nurse and/or social worker present.

(2) Circumstances that in the opinion of the team merit special attention from a psychiatric viewpoint should involve a *routine* psychiatric referral. In our clinic a routine investigation is prescribed (a) for all couples considered for AID, to assess the couple's motivation, the stability of the marriage, and the capacity for parenthood, and to discuss some of the legal difficulties; and (b) after the infertility work-up for every so-called physically 'normal couple'.

(3) Every male who has been found to have abnormal findings on semen analysis should have a follow-up with special attention to sexual performance.

(4) Because marital problems often become manifest during an infertility work-up, every reproductive biology unit should have available a facility that deals with such problems both with short-term supportive and behaviorally oriented techniques and with more intensive therapies.

(5) *Every* staff member of a reproductive biology unit should become involved and familiar with psychological, ethical and legal issues pertaining to infertility. This can best be accomplished by an ongoing conference in which such issues are discussed. The sticky question of whether a couple can offer adequate parenting for a child for example, or the hazards of pregnancy when the parents are over 40, a situation which requires genetic counselling, should not be dealt with by the psychiatrist alone.

(6) After completion of the work-up couples should be encouraged to feel free to consult with the nurse, social worker or psychiatrist at a later date, if they so wish, in regard to psychological problems.

Summary

The psychological aspects of male infertility cannot be isolated from issues pertaining to the couple as a unit or to the clinical setting. The work-up will in itself be traumatic. Marital problems may masquerade as an infertility problem. Psychological factors may cause infertility via their effect on sexual performance. That emotional conflicts can affect fertility directly via physiological pathways has not been established. The couple also requires help in dealing with the outcome of the work-up. How a reproductive biology clinic might utilize a psychiatrist is discussed.

References

1 BERGER, D.M.: Psychological assessment of the infertile couple. Can. Family Physician 20: 89–90 (1974).
2 BULLOCK, J.L.: Iatrogenic impotence in an infertility clinic: illustrative case. Am. J. Obstet. Gynec. 120: 476–478 (1974).
3 CURTIS, J.L.: A psychiatric study of 55 expectant fathers. U.S. arm. Forces med. J. 6: 937–950 (1955).
4 DUBIN, L. and AMELAR, R.D.: Sexual causes of male infertility. Fert. Steril. 23: 579–582 (1972).
5 HYYPPA, M.; RINNE, U.K., and SONNINEN, V.: The activating effect of L-dopa treatment on sexual functions and its experimental background. Acta neurol. scand. 43: suppl. 46, p. 223 (1970).
6 KREUZ, L.E.; ROSE, R.M., and JENNINGS, J.R.: Suppression of plasma testosterone levels and psychological stress. Archs gen. Psychiat. 26: 479–482 (1972).
7 MALMNAS, C.O. and MEYERSON, B.J.: Monoamines and copulatory activity in the castrated male rat. Abstract. Acta pharmac. tox. 31: suppl. p. 1 (1972).
8 MASTERS, W.H. and JOHNSON, V.E.: Principles of the new sex therapy. Am. J. Psychiat. 133: 548–554 (1976).
9 MEHAN, D.: Clinical application of meiotic preparations in the infertile male. Presented at 32nd Ann. Meet. Am. Fertility Soc., Las Vegas 1976.
10 PLATT, J.J.; FICHER, L., and SILVER, M.J.: Infertile couples: personality traits and self-ideal concept discrepancies. Fert. Steril. 24: 972–976 (1973).

D.M. BERGER, MD, Mount Sinai Hospital, Department of Psychiatry, 600 University Avenue, *Toronto, M5G 1X5* (Canada)

Newer Concepts in Marital and Sex Therapy

David L. Shaul

Mount Sinai Hospital, Division of Psychosomatic Obstetrics and Gynaecology, Toronto, Ont.

Changing Attitudes

One of the most important features which has come to light in understanding sexual disorders of the male is the changing attitudes of society and the changing roles of the male and female in our society. Previously, the male played the dominant role in sex. If the male was satisfied, there were no problems. It was felt that he was not called upon to consider his partner whatsoever – her feelings, her desires, or her sexual satisfaction. Today, the female is becoming more aware of her sexuality. 'Over the centuries', say Masters and Johnson, 'the single constant etiologic source of all forms of male sexual dysfunction has been the level of cultural demand for effectiveness of male sexual performance. Women need only to lie still to be potent.'

The Role of Culture

The couple who abides strictly by the rules of its religion, especially that religion which demands absolute virginity at the time of marriage, and that which teaches that sex other than for procreation is sinful, will probably have problems in the sexual area. A religious ceremony cannot suddenly make them comfortable and lose themselves in the sexual act.

Preparation for Role as a Counsellor

Physicians are in a unique position to offer patients advice and counsel, but training and preparation for the role of counsellor has often been inadequate. Many times, couples complaining of sexual dysfunction first come to their physician as the logical source of information and help. If the physician is properly prepared, and if he has dealt with his own prejudices and come to terms with his own sexuality, he has a great opportunity to render an extremely important service.

It is important for the therapist to be secure in his own right. If he does not feel that a case is within his realm, he should be secure enough to call on consultants from among other medical or paramedical personnel.

A physician may develop a sensitivity to people's needs which acts as an antenna to alert him to the true message the client is giving him. When this is developed, a line of communication is opened which eventually will bring out the client's real reasons for wanting help. This can only be developed when the professional is willing and able to see these signals and respond to them. A case which illustrates this point follows:

Mrs. X was a 52-year-old woman who 4 years previously had a hysterectomy performed for large fibroids. On each of her annual visits she was always well-groomed and well-mannered. On a recent visit her manner seemed the same as always but it was noted that her hair was slightly dishevelled in that one curl was out of place. In addition her blouse had one button undone. This seemed out of character for her, although her voice and manner appeared unchanged. The open button on her blouse and the displaced curl on her hair were an invitation to come into her life and ask more. When this was done she openly expressed concerns which she had kept hidden for many years. As a result of this interview she unfolded a story which indicated her need for further help. She was referred to the appropriate agencies and subsequently found comfort and solace in her more complete therapy.

Anatomy and Physiology of Sexual Response

MASTERS and JOHNSON [5] divide the responses of both sexes into four stages: (1) excitement; (2) plateau; (3) orgasm, and (4) resolution. The response in the reproductive organs can be described as follows.

Excitement

There is sexual arousal, with erection of the penis in the male, and vaginal lubrication in the female.

Plateau (fig. 1)

A more advanced stage of arousal which occurs immediately prior to orgasm. This is the peak of vasocongestion that occurs in both sexes. In the male, the penis and testicles are filled with blood; in the female, the vaso-

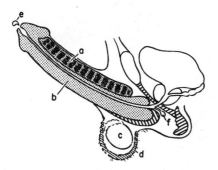

Fig. 1. The male genitals in a highly aroused state (plateau). The corpora cavernosa (a) and the corpus spongiosum (b) are filled with blood, causing erection of the penis. The testicles (c) are also engorged and increase in size and just before orgasm rise against the perineal floor. The dartos tunic (d) which covers the testes is thickened and contracted. A drop of clear mucoid secretion (e) from Cowper's gland (f) appears at the urethral meatus during intense excitement. (Reprinted by permission of author H.S. KAPLAN).

congestion gives rise to the sex skin, the orgasmic platform, and other manifestations of swelling and coloration of the labia minora and the tissue around the entrance to the vagina. In the female, lubrication is an important feature of this stage.

Orgasm (fig. 2)

In the male, semen squirts out of the erect penis at 0.8-sec intervals. In the female a similar type of rhythmic contraction of the circumvaginal and perineal muscles and the orgasmic platform occurs. After orgasm the male is refractory to sexual stimulation, whereas the female is not. This accounts for the ability of the female to have multiple orgasms.

Resolution (fig. 3)

The above described responses abate, and the body returns to its basal state.

More recently, PION *et al.* [7] in Honolulu have pointed out the advantages in therapy of realizing the similarity in the responses of the male and the female. From an anatomic and physiologic standpoint, the two components of sexual response consist of the following:

(1) The vasocongestive reaction which causes erection in the male and lubrication in the female. These are under control of the parasympathetic nervous system.

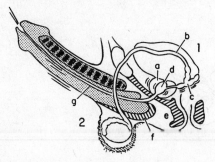

Fig. 2. The male genitals during orgasm. Phase 1 – emission: This phase is perceived as the sensation of 'ejaculatory inevitability'. The internal male reproductive viscera – prostate (a), vas deferens (b), seminal vesicles (c) – contract and collect the ejaculate in the urethral bulb (d). Phase 2 – expulsion: The perineal (e), and bulbocavernosus (f) muscles contract with a 0.8-sec rhythm causing pulsations of the penis and explusion of the ejaculate. The penile urethra (g) contracts also. (Reprinted by permission of author H.S. KAPLAN).

(2) The muscular contraction which is ejaculation or orgasm in the male, and orgasm in the female. These are under the control of the sympathetic system.

Since there are two controls, each of a different component of the response, so can there be a different clinical syndrome due to dysfunction of these different controls.

Approach to the Client

Much has been written about the value of single therapy or cotherapy. In dealing with a female client, a male therapist and a husband might cause the patient to feel that she is being outnumbered by the males two to one, or conversely, the husband sitting with a female therapist and his wife might feel the same way. Both methods are used in our center and in many instances it is found that single therapy is equally as productive. It is always expressed to the patient that the presence of both partners would be more desirable. However, suggestions in therapy can be made in three different ways: (1) to the male alone; (2) to the female regarding the male, and (3) to the couple, which is best, when both are willing to come. MASTERS and JOHNSON [6] emphasize that there is no such entity as an uninvolved partner

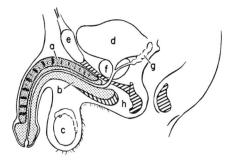

Fig. 3. The male genitals in a quiescent state. The penis is flaccid because there is relatively little blood in the corpora cavernosa (a) and in the corpus spongiosum (b). The testes (c) are in their normal low position during quiescence. The drawing further represents the urinary bladder (d) and its anatomic relationships to the pubic bone (e), the prostate (f) and the seminal vesicles (g), and a schematic representation of the bulbocavernosus and perineal muscles (h). (Reprinted by permission of author H.S. KAPLAN).

in a marriage contending with any form of sexual inadequacy. Sexual dysfunction is a marital unit problem, not a husband or wife problem. We have treated a single person but only on the following terms (1) that there is a sexual partner; (2) that the presence of the sexual partner would add to the therapy; (3) anyone will be seen for at least one visit to do an assessment and screening; (4) the door is always open for singles should they want to return when they again have a partner.

SAGER [10] states that marital and sex therapy are currently interrelated in most cases, because they deal with different symptoms of overlapping aspects of the couple's total relationship. The connection between marital discord and sexual dysfunction determines the emphasis and course of treatment.

For example, he states that sex therapy is the immediate treatment of choice when sexual dysfunction produces secondary marital discord, but not when severe discord precludes the possibility of good sexual function. In his series, 75% of the couples had a mixture of marital and sexual problems in varying proportions. Our results are similar. FRANK *et al.* [2] studied profiles of couples seeking sex therapy and marital therapy. Their findings tended to confirm what has been most often found – the relationships of the couples undergoing primary sex therapy were generally characterized by satisfaction and affection, whereas those of the couples undergoing marital therapy were often antagonistic.

Very often on the initial visit one can determine whether or not a couple is suitable for sexual dysfunction therapy. The counsellor must be prepared to realize his own limitations, and to make the necessary referrals when these have been reached. In some cases, concurrent therapy with the appropriate agency can be carried on, e.g. intensive psychotherapy of one partner of the couple does not exclude dealing with a problem of sexual dysfunction of the couple as a unit.

Sex-Oriented History

Various methods of taking a history have been described. In MASTERS and JOHNSON's [6] book an entire chart is presented. We use a medical history, and orient it to sex problems, as suggested by the work of ANNON [1].

The outline of the history is divided into five parts as follows:

Part 1	A description of the current problem
Part 2	Onset and course of problem A. Onset (gradual or sudden; precipitating events, consequences) B. Course (changes over time: increase, decrease, or fluctuations in severity, frequency, intensity; functional relationships with other variables)
Part 3	Client's concept of cause and maintenance of the problem
Part 4	Past treatment and outcome A. Medical evaluation (speciality) B. Form of treatment (results, currently on any medication for any reason) C. Other professional help (speciality, date, form of treatment, results) D. Self-treatment (type and results)
Part 5	Current expectations and goals of treatment (concrete versus general)

In taking this history, it is extremely important that the therapist listen very carefully to what the client is saying. ANNON [1] alludes to the three L's: Listening, Labelling, and Language. He shows the importance of understanding what the client is saying, of avoiding putting labels or patterns on this, and also that the client understands what we are saying. Although the history is structured for the therapist, it does not appear so to the client.

Examination

Before starting therapy, the client has a complete physical examination, to rule out any medical conditions which can affect his response. For example, certain chronic debilitating diseases, such as diabetes and multiple sclerosis, and other diseases affecting the nervous system can offer physical factors which would affect the client's response. An adequate physical examination and appropriate laboratory work are indicated on both clients before undertaking this type of work.

Definition of Sexual Dysfunction in the Male

The specific complaints with which males present may be divided into three categories: (1) premature ejaculation; (2) ejaculatory incompetence, primary or secondary, and (3) impotence, primary or secondary.

Premature Ejaculation

Premature ejaculation is a difficult term to define. MASTERS and JOHNSON [6] define a premature ejaculator as a man who is unable to control ejaculation sufficiently well to satisfy his partner on at least 50% of the occasions when intercourse is attempted. It should be noted that they have not orgasm a requirement of female satisfaction in their definition. Premature ejaculation carries a high success rate in therapy. The causes can usually be determined from the pattern of early sexual experience – the need to have a rapid orgasm for fear of discovery in a parked car or a living room couch; the protestations of the prostitute not to take too long so that she can get on with her next client, etc. A disinterested woman who asks her partner to hurry can perpetuate this situation. 75% of males ejaculate in less than two minutes. Ejaculation is faster after periods of abstinence. As the male attempts to delay orgasm by thoughts, condoms and creams, he becomes a spectator to his performance rather than an involved partner in a pleasurable process. This can start a vicious circle, and can lead to secondary impotence.

Ejaculatory Incompetence

This is the reverse of premature ejaculation. It can be primary or secondary in character. A man with ejaculatory incompetence rarely has difficulty in achieving or maintaining an erection sufficient for successful coitus but he is unable to ejaculate and experience orgasm. An anxious male with this disability, like the premature ejaculator, can develop secondary impotence. MASTERS and JOHNSON [6] report on 17 cases. The factors in such a small number are difficult to pinpoint. Some mentioned are: (1) strict upbringing; (2) dislike, rejection, or open emnity for their wives; (3) fear of pregnancy; (4) maternal dominance; (5) homosexual preference; (6) specific instances in a previously satisfactory sexual response – (a) a husband finding his wife while she is committing adultery; (b) interruption during intercourse by their children.

Impotence

Erection either cannot be achieved or cannot be maintained long enough to permit penetration of the vagina. Impotence, or erectile failure, may be either primary or secondary. There are many possible causes, but the majority of cases are due to emotional or psychological factors. There is often a specific event that sets it off – guilt, drugs, jet lag. A man cannot usually be highly anxious and have an erection at the same time. Once erectile failure has occurred, one must look for the factors that maintain the problem and continue his anxiety.

General Concepts of Therapy

Many types of therapeutic approaches are possible. HARPER [3] described 36 separate systems of psychotherapy. Others have been added since. Different approaches to therapy will depend upon the discipline of the therapist; i.e., psychologist, psychiatrist, family practitioner, obstetrician and gynecologist, urologist, endocrinologist, nurse, or social worker. The modes of therapy available range from the 2-week rapid treatment of dual sex therapy of MASTERS and JOHNSON [6], the new sex therapy of KAPLAN [4] to the behavioral treatment of sexual problems by ANNON [1].

These, coupled with personal communications with PION *et al.* [7] have been used by us to establish our format. Patients may be seen alone or with a cotherapist who is either a social worker or a nurse counsellor. As stated previously, we prefer to see the couple, but will see a single member of the unit, or a single person, on an assessment basis for screening purposes. Medical evaluation of all patients has previously been carried out.

Since this is a brief encounter format, the interview proceeds as quickly or as slowly as necessary to elicit helpful information. The goals and aspirations have already been covered in the sex-oriented history. The couple is not given the impression that they are on trial. The treatment is directed towards the main complaint. The model of treatment which we have found most successful is the P-LI-SS-IT model as described by ANNON [1]. It provides for four levels of approach. These are:

(1) P – Permission
(2) L – Limited
    I – Information
(3) S – Specific
    S – Suggestions
(4) I – Intensive
    T – Therapy

The first three levels can be viewed as brief therapy. Each succeeding level requires more time and more skill. By using this model the therapist knows how deeply he is becoming involved, and whether he feels comfortable in going further. At any level he can decide he has reached the endpoint of his capability, and may elect to refer the client to the appropriate agency. This requires the therapist to have a backup team of consultants to take over in such a situation, and not leave the client hanging in a vacuum. There are general concepts of therapy which apply to all male sexual dysfunction. These can be applied to couples, no matter what the problem. It is important to emphasize that it is the situation that is under study and that it is the specific dysfunction that is to be corrected rather than faults in either the husband or wife. The female's cooperation in all of these events is essential. KAPLAN [4] points out that many times the wife and her attitude in the relationship between the two can be a destructive factor in attempting to help the husband.

Having taken an adequate sexual history, many times permission and limited information may be enough to effect a satisfactory result. Anxiety having been allayed, the couple is on the way to recovery. In these cases it was not what they were doing, but their concern about what they were doing that was the cause of their problem.

When permission and limited information are not enough to attain the goal for which the couple and the therapist have set themselves, certain types of specific suggestions can then be put forward. These include (1) nongenital stimulation, (2) genital stimulation, (3) nondemand coitus, and (4) finally sexual activities as defined in the goals of therapy.

Nongenital stimulation: In taking the sexual history, one attempts to determine whether or not the couple did indeed have pleasurable experiences before. Many times it can be elicited that they did. Most commonly we find these experiences were in a nondemand situation, such as a couple who, in premarital pleasuring had agreed that they would not have sexual intercourse. These can be remembered by the couple as pleasurable experiences and can be used in therapy. The first aim is to remove all pressure of performance. Nonsexual tasks are given with the requirement that no sexual intercourse will be undertaken. They may touch each other, they may caress each other and in some instances it is found helpful to advise them to take turns in massaging each other with various creams and lotions. Anything that gives them pleasure is advised but no genital stimulation, and no sexual intercourse are allowed. They are sent home with these tasks. In certain cases we can use male and female dating, where one night is his night, where he determines what shall be the activities for that night – what type of a dinner he would like and what type of activity he would like to engage in which would give him good feelings. The female gets her turn as well. When this first task has been satisfactorily completed, they move on to genital stimulation. The male is asked to think of what he finds most pleasurable and to tell his wife what she can do for him. He is advised that it is not wrong to have erotic thoughts and fantasies. It is helpful, and creates a greater bond of affection if the husband directs his wife's actions by holding her hand in his and helping her to please him.

Once this has been learned, more specific tasks relating to the complaint are given.

Case 1

This is the case of a 30-year-old male who had been married 2½ years. He and his wife had no children. His chief complaints were: (1) I can have an orgasm but not during intercourse; (2) sex: I can take it or leave it. In the past he had seen a psychiatrist whom he felt had her own problems. He found the current visit painful but nonetheless he was able to suggest possible causes of his presenting complaints: (1) his wife had to take a shower before intercourse (2) she was a secretary for a charitable organization which had out of town committee meetings and she did not get home until late at night (3) he was a teacher and he had to stay up until 2 or 3 in the morning marking papers so that he could keep up with his lessons. For 1 year prior to their marriage they lived together in their own apartment out of the country but had never had intercourse. They masturbated each other then and this is what they continued to do until the present. They were given nonsexual pleasuring tasks and told not to have intercourse for the time being. They were also instructed to have more direct verbal communication with each other. At first the wife enjoyed 'the exercises' fully but the husband did not like the touching.

He was 'ticklish'. With time this passed and his erections which used to be lost prematurely, were now maintained. They were given permission for genital pleasuring and vaginal intercourse. Within 2 weeks they reported successful intravaginal penetrations and orgasm. On their final visit they stated that the showers, committee meetings and the school lessons were no longer problems. Within 3 months of leaving treatment, the wife became pregnant for the first time. Subsequently both husband and wife appeared together for prenatal visits.

Application of Treatment to Specific Complaints

Premature Ejaculation

Once the premature ejaculator has been able to get pleasure from nongenital stimulation, he is ready for his next set of tasks. The husband and wife are now allowed to engage in genital stimulation. There are different methods to delay ejaculation.

(1) Redirection of attention [7]. The male lies on his back and relaxes his buttock muscles. Husband and wife try different movements. He tries to think what he feels in his penis and what type of feelings he is having. They see which movements hurry and which delay an orgasm.

(2) SEMAN's [11] technique. When the male feels he may ejaculate his partner stops her caressing. When the feeling passes, and even if the erection is lost, caressing is restarted. This stopping and starting gives the male confidence that his ejaculation can be delayed.

(3) MASTERS and JOHNSON [6] squeeze technique. When the male feels the urge to ejaculate his partner applies pressure to his penis with her thumb and first finger for 3–4 sec. The male will respond to sufficient pressure by losing his urge to ejaculate and some of his erection. The wife should allow an interval of 15–30 sec after releasing the applied pressure and then return to active penile stimulation. Again, when full erection is achieved, the squeeze technique should be reinstituted. Alternating between periods of pressure and of sexually stimulative techniques may give up to 15 or 20 min of sex play before ejaculation occurs.

(4) Increased frequency of ejaculation usually with masturbation, in the opinion of PION *et al.* [7] will eventually delay orgasm. PION suggests four to five times in a weekend. The male does not need any new techniques and can just continue with this method. Pressure to have orgasm with a partner is removed. When orgasms do occur he begins to have more positive feelings and because he has them more frequently, his erection stays longer and his orgasm is delayed.

The male now attempts to enter the vagina. The couple is given instruction in nondemand intercourse with the female superior position. She lies perfectly still, and he determines how much thrusting he will do. He can take as long as he wants. If he should feel the urge to ejaculate prematurely, he instructs his partner to remove herself. With repeated genital stimulation and repeated vaginal entries in this manner, ejaculation can be delayed. This delay is prolonged with the methods described above.

Ejaculatory Incompetence

MASTERS and JOHNSON [6] found only 17 cases with this complaint. The possible causes have been alluded to. In this situation a stronger type of erotic stimulation or masturbatory situation may be necessary in order to help these males achieve their orgasms. One must remove the need to perform from the male. Nonsexual tasks are started as with premature ejaculation, and as these become more erotic and exciting, extravaginal stimulation is carried out. When orgasm occurs in a satisfactory time, this orgastic stimulation can be used with vaginal containment.

Case 2

The wife, aged 24, husband aged 27, were married 3 years and had no children. They had been seeing another physician for 10 months for various complaints. Neither one had a satisfactory orgasm with vaginal intercourse. Although he had no erectile difficulty, the husband took 1 h to reach orgasm within the vagina. She could not have an orgasm with vaginal containment but only by oral or digital manipulation. She became frustrated because of his delayed orgasm. They were having difficulty in achieving pregnancy because of the above complaints.

Their past history revealed that they courted for 3 years, during which time she did not permit vaginal intercourse. They were married 2½ years before she allowed vaginal intercourse. As stated, oral and digital sex were enjoyed. She was afraid of vaginal intercourse and he, because he wanted to be faithful and kind, did not complain. Now that they wanted to have a child, they both had to come to grips with the problem of vaginal intercourse.

She had gone to her physician for 10 months for the treatment of vaginismus. Her vagina was first dilated by her physician, then by herself, and now by her husband. She was quite pleased with what had happened to her, but her husband suddenly found that he could not have an ejaculation in less than an hour. Because they both seemed to have a problem, they were instructed in husband and wife pleasuring. They were given nonsexual tasks but they were not to have intercourse. They were to enjoy each other in the activities they were doing. 3 weeks later they returned and reported success. On their own they had progressed to genital stimulation and he could have an orgasm in about 5 min. They looked forward with anticipation to progressing further with vaginal intercourse.

In this particular instance the male's difficulty was due to the fact that his wife, for 3 years in their courtship, and in the first 2½ years of their marriage would not allow him to have vaginal intercourse because of her fear of being damaged. The stimulations which he found very highly orgastic were repeated, as he lay on his back being brought close to an orgasm. At this point, stimulation was continued as she took the female superior position and inserted his penis into the vagina. Once in this position, she continued the stimulation with her hand, and eventually he had his first orgasm with vaginal containment. Once this occurred, further progress was rapidly made.

Impotence

Treatment of impotence is carried out in the same way as for the premature ejaculator. Nongenital tasks first, with progression to genital stimulation, nondemand sex, and finally the desired sexual activity are employed. The onset of the erectile difficulty can often be specifically determined. This should be elucidated if at all possible. Once this factor comes out, it is discussed, and its role in the couple's problem is pointed out. This often results in a gradual lessening of tension between the couple and a gradual lessening of fear in the male that he will no longer be able to perform. It is essential to rule out possible medical conditions as causative factors.

Case 3
A 65-year-old man was married for almost 40 years. He complained that where he once could hold an erection for 30–50 min, he now had an ejaculation or loss of erection immediately before or just after entry. He stated that he loved his wife but they were not as close as they used to be. This complaint dated back to a severe depression from which he had recovered 1½ years previously. He was advised to return with his wife and on subsequent visits their lack of communication became evident. His need to perform and to succeed and fear of failure permeated the discussion. During the first two visits the discussions were very much the same with little progress being made. On a subsequent visit he stated that he had had no erections, where in truth he had had 5–6 erections, three with orgasm in the previous 4 weeks. When questioned about this he stated he did not consider these relevant because they were not obtained with vaginal intercourse. Also for the first time he began to complain about the fact that his wife would not undress in front of him and this was something which he felt could make him more stimulated. She was visibly shaken by his statement and had a tearful reaction. She stated that at her age she did not feel that she should have to change her habits, i.e. the undressing especially upset her. The two of them actually were very surprised at the lack of knowledge each had about the others needs. When they could openly talk about these they both realized how deeply this affected them. They resolved to go home to discuss it further.

On a visit 2 weeks later, they were both much more relaxed. The discussion was much more penetrating and his anxiety about his age, his ability to continue his professional work and some of the bad relationships he was having in business and with his family came out. He stated that he now understood that the anxiety was operative in his impotence and it became apparent that his impotence was not only in his sexuality but also in his inability to control his future. He and his wife were quite excited about finally understanding the nature of their problem. From this point on he began to improve rapidly.

Other Therapeutic Modes

Surgery

PRYOR [8] states that 'surgery has a small but important role in correcting sexual disorders of men. Some of the symptoms requiring surgical correction are well recognized, others are less well known and are often ignored.'

Painful coitus. The tight foreskin which causes a painful erection is readily recognizable and easily cured by circumcision. Certain injuries to the penis may cause painful coitus: (a) hyperflexion of the erect penis with ruptured tunica albuginea; (b) pain in the penile shaft due to Peyronie's disease, and (c) a perplexing symptom is one of pain at the tip of the penis at the time of ejaculation. Although the cause is often difficult to determine, prostatitis should be ruled out.

Penile deformity. Ventral curvature of the penis is a feature of hypospadias associated with chordee. Ventral curvature of the penis may also be due to chordee without hypospadias. Surgical treatment lies between lengthening the urethra and cutting the chordee. Dorsal curvature of the erect penis is a feature of Peyronie's disease and the angulation is often severe enough to prevent penetration of the vagina. In many instances this disease will correct itself spontaneously. Surgical excision of the fibrous tissue is only indicated when prolonged medical treatment fails to improve the deformity.

Priapism. The initial satisfaction produced by long sustained erection soon passes to embarrassment and finally fear. Medical treatment is rarely sought during the first 12 h. Conservative management is indicated only while the patient is being prepared for surgery. Sickle cell disease and leukemia must be ruled out. The treatment is surgical decompression of the engorged corpora cavernosa.

Sex Aids

A variety of physical and chemical aids to enhance sexual pleasure are currently available. These include vibrators, assorted clothing items, creams, lotions, perfumes, etc.

RHODES [9] indicates that the use of these aids is still controversial but feels that the medical profession should be aware of them. He believes that if certain sexual problems can be at least in part resolved by use of these aids, they deserve a trial.

In an unselected random group of couples coming for sexual dysfunction therapy, an 85% improvement rate can be expected for couples not requiring intensive psychotherapy. This can often be effected in just a few visits using the behavior modification approach described above.

References

1 ANNON, J.S.: The behavioural treatment of sexual problems, vol. 1: Brief therapy, pp. 56–58, 100 (Enabling Systems, Honolulu 1975).
2 FRANK, E.; ANDERSON, C., and KUPFER, D.J.: Profiles of couples seeking sex therapy and marital therapy. Am. J. Psychiat. *133:* 559–562 (1976).
3 HARPER, R.A.: Psychoanalysis and psychiatry: 36 systems (Prentice Hall, Englewood Cliffs 1959).
4 KAPLAN, H.S.: The new sex therapy, pp. 13 (Brunner/Mazel, New York 1974).
5 MASTERS, W.H. and JOHNSON, V.E.: Human sexual response, pp. 3–8, 68, 177 (Little, Brown, Boston 1966).
6 MASTERS, W.H. and JOHNSON, V.E.: Human sexual inadequacy, pp. 3, 34–51, 103, 117 (Little, Brown, Boston 1970).
7 PION, R.J.; ANNON, J.; ROBINSON, C.; REICH, L., and PION, G.: Personal commun. (1975).
8 PRYOR, J.P.: Surgery of male sexual disorders. Br. med. J. *iii:* 585–587 (1975).
9 RHODES, P.: Sex aids. Br. med. J. *iii:* 93–95 (1975).
10 SAGER, C.J.: The role of sex therapy in marital therapy. Am. J. Psychiat. *113: 5:* 555–558 (1976).
11 SEMANS, J.H.: Premature ejaculation: a new approach. Sth. med. J. *49:* 353–362 (1956).

Dr. D.L. SHAUL, 123 Edward Street, *Toronto, Ont. M5G 1E2* (Canada)

Contributors

Rudi Ansbacher, Department of Obstetrics and Gynecology, Brooke Army Medical Center, Fort Sam Houston, Tex., USA.

Jerald Bain, Departments of Medicine and Obstetrics and Gynecology, University of Toronto and Mount Sinai Hospital, Toronto, Ont., Canada.

B. Norman Barwin, Department of Obstetrics and Gynecology, University of Ottawa and Ottawa Civic Hospital, Ottawa, Ont., Canada.

David Berger, Department of Psychiatry, University of Toronto and Mount Sinai Hospital, Toronto, Ont., Canada.

Bruno Dahlberg, Department of Obstetrics and Gynecology, University of Lund, Malmö, Sweden.

E.S.E. Hafez, Reproductive Physiology Laboratories and Andrology Research Unit, C.S. Mott Center for Human Growth and Development, Department of Obstetrics and Gynecology, Wayne State University School of Medicine, Detroit, Mich., USA.

Robert W. Hudson, Department of Medicine, University of Ottawa and Ottawa General Hospital, Ottawa, Ont., Canada.

Elaine E. Jolly, Department of Obstetrics and Gynecology, University of Ottawa and Ottawa Gneral Hospital, Ottawa, Ont., Canada.

Philip G. Klotz, Division of Urology, Department of Surgery, University of Toronto and Mount Sinai Hospital, Toronto, Ont., Canada.

David E. McKay, Division of Urology, Department of Surgery, University of Ottawa and Ottawa General Hospital, Ottawa, Ont., Canada.

Murray Miskin, Department of Radiological Sciences, University of Toronto and Mount Sinai Hospital, Toronto, Ont., Canada.

Frederick Naftolin, Department of Obstetrics and Gynecology, McGill University and Royal Victoria Hospital, Montreal, Que., Canada.

R.F. Parrish, Department of Obstetrics and Gynecology, Washington University School of Medicine, St. Louis, Mo., USA.

Kenneth L. Polakoski, Department of Obstetrics and Gynecology, Washington University School of Medicine, St. Louis, Mo., USA.

David Shaul, Division of Psychosomatics, Department of Obstetrics and Gynecology, University of Toronto and Mount Sinai Hospital, Toronto, Ont., Canada.

M.H.K. Shokeir, Division of Medical Genetics, Department of Pediatrics, University of Saskatchewan and University Hospital, Saskatoon, Sask., Canada.

L.J.D. Zaneveld, Department of Physiology and Biophysics, University of Illinois at the Medical Center, Chicago, Ill., USA.

Subject Index

Abstinence 8, 13, 14
Accessory sexual glands 9, 13
Achondroplasia 86
Acid phosphatase in seminal plasma 29
Agglutination of spermatozoa 50, 51, 55, 57, 61
AID 141, 147–150, 162, 163
AIH 19, 141, 143–146
Ambiguous genitalia 79–81
Androgens, production 4
Anti-estrogens 136
Arginine, stimulation of sperm 25
Artificial insemination 66, 105, 106, 114, 141–150
– –, technique 149, 150
Ascorbic acid 66
Asthenospermia, treatment 135
Autosomal disorders 83–89
– –, dominant 85, 86
– –, recessive 86–91
Azoospermia 23, 106
–, treatment 113

B lymphocytes 60
Bacteriospermia 47–49
–, asymptomatic 48–52, 58
–, treatment 50
Balanced polymorphism 97
Barr bodies 73, 78
Blood testis barrier 3, 62
Bouins solution 108
Buccal smear 74, 80

Caffeine, stimulation of sperm motility 25
Capacitation 56
Catecholamine 103
Candida albicans 54, 55
Cervical mucus 63–66, 110
Chlamydia trachomatis 53, 55, 111
Citric acid in seminal plasma 28
Clomiphene citrate therapy 136
Coagulation and liquification of semen 17–19
Coitus interruptus 15, 111
Congenital adrenal hyperplasia 88, 89
– anorchia 87
Contraception 128–130
Cortisol 112
Cowper's gland 12
Cri-du-chat syndrome 84
Cryptorchidism 6, 95, 104, 105
Cyclic AMP 42
Cystic fibrosis 86, 97
Cytotoxic antibody 61, 64

Del Castillo syndrome 84, 95
Diabetes mellitus 95
Dihydrotestosterone (DHT) 38–40
–, seminal plasma 39, 40, 43
DNA 1
L-Dopa 161
Down's syndrome 84, 95
Ductus deferens, *see* Vas deferens

Ectopic testis 6

Subject Index

Efferent ductules 6
Ejaculation 168
Ejaculatory duct 7
– incompetence 171
Endoscopy 80
Epididymitis 104, 125, 126
Epididymo-vasostomy 109, 113, 114
Epididymus 6–8, 108, 109
Escherichia coli 47, 54
Estrogen 40, 41
Eunuchoidism 88, 132

Familial hypogonadotrophic eunuchoidism 88
Follicle stimulating hormone (FSH) 2, 33–38, 42, 43
Fructose in seminal plasma 28

Germinal aplasia 35
Gonadal biopsy 81
– dysgenesis, XY female type 94
Gonadotropin(s) 33, 36, 43
–, therapy 135
– releasing hormone (GnRH) 33–36, 136
Grey scale 118–120
Gynecomastia 73

Hemospermia 20
Human chorionic gonadotrophin (HCG) 104, 135
H-Y antigen 82
Hyaluronidase 63
Hydrocele 123–125
Hyperspermia 20
Hypertelorism 87
Hypogonadotrophic hypogonadism 135
Hypospadias 95, 96, 105, 178
Hypospermia 20
Hypothalamic disease 133
Hypothalamus 9
Hypothyroidism 95, 138

IgA 60, 63
IgD 60
IgE 60
IgG 60, 61, 63
IgM 60, 61, 64

Immobilizing antigen 63
Immunofluorescence 62
Impotence 157, 159, 172
–, treatment 177
Inhibin 33, 42, 43
Intertubular tissue 3
Isoimmunization 61

Kallikrein, stimulation of sperm motility 26
Kallmann syndrome 93, 135
Kinocilia 3
Klinefelter syndrome 71–74, 82, 113, 120, 121, 131, 132

Lactobacillus 54
Laparoscopy 81
Laparotomy 81
LDH 63
Leydig cell 4, 34, 36–38, 133
Littre's gland 9, 12
Live-dead (supravital) stain 24, 25
Lumen 3, 6
Luminal epithelium 8
Luteinizing hormone (LH) 33–38, 42, 43, 133, 135–138, 148
– – releasing hormone (LHRH) 43, 136, 137
Lyonization 94

Masturbation 14, 160, 175
Meningomyelocele (spina bifida) 96
Mesterolone 138
Monoamines 161
Monoamino acid oxidase antagonists 104
Morphology of sperm 26, 27
Motility of sperm 23–25, 39, 40, 106, 141
– – –, stimulation 25, 26
– – –, bacteria and 47,, 52, 53
Mullerian duct syndrome 88
Mumps 105
Mycoplasma 54, 55, 111, 135
– *hominis* 48, 55
Myotonic dystrophy 85, 86
Myoid cells 3

Neoplasm 212
Nitrofurantoin 103

Subject Index

Noonan's syndrome 85
Normospermia 23, 34–41

Oligozoospermia 23, 34–41, 43, 106
–, treatment 111, 134–138
Orchitis 121
–, allergic 60
Orgasm 167
Ovitestis 81
Oxytocin 8

Pampiniform plexus 6
Pargyline hydrochloride 104
Parkinsonism 161
Penis 166–169
–, deformity 178
Peyronie's disease 178
pH, seminal fluid 21
– – – and bacteriospermia 58
Phenelzine sulphate 104
Pituitary-adrenal axis 161
– disease 133
– tumour 135
Polygenic disorders 94–96
– –, generalized 95
– –, localized 95, 96
Polyzoospermia 23
Premature ejaculation 160, 171
– –, treatment 175, 176
Priapism 178
Prolactin 40, 41
Prostate 9, 13, 111
Prostatitis 53, 104, 111
Prostatovesiculitis 52, 53
Protamine 63
Prozone effect 62
Pseudohermaphroditism 93
Pseudovaginal perineoscrotal hypospadias 87
Puberty 3
–, precocious 3

Reifenstein syndrome 83, 93
Reiter's syndrome 53
Rete testis 120
Robinow syndrome 95

Salpingitis 55
Scrotum 4
Semen 12
– analysis 105, 106
– –, biochemical 27–29
– collection 14, 15
– and infections 55–58
– preservation 150–153
– –, technique 17–25
Seminal plasma 12
– androgens 39
– vesicles 9, 12, 108, 111
– vesiculography 108, 109
Seminiferous tubules 3, 33, 36
Seminoma 122
Sertoli cell only syndrome 94, 113
– cells 4, 5, 34, 42
Sex aids 178, 179
– reversal syndrome 86
Smith-Lemli-Opitz syndrome 95
Sperm antibody 62–66, 110
– –, tests 64
– antigens 62–66
– longevity 49, 55, 56
– viability and infections 52, 53
Spermacides 22
Spermatocele 104, 125, 126
Spermatogenic function 1–3, 132, 133
– –, stress and 103
– –, testicular biopsy and 107
Spermatogonial division 1
Spermatozoa 12, 62
–, correlation with testis size 126, 127
–, count and concentration 22, 23, 106
–, penetration of cervical mucus 29, 30, 61, 65
Spermine 20
Spina bifida 96
Staphylococcus albus 47, 49
Sugar stimulation of sperm motility 25

T lymphocytes 60
T-mycoplasma 55, 111
Tamoxifen 137
Telecanthos-hypospadias syndrome 83
Testicular abscess 121–123
– biopsy 37, 106–108, 133

Subject Index

Testicular abscess
– capsule 3
– feminization 91
– –, incomplete 92
– hematoma 121
– ischemia 3
– tortion 104
Testis 1–10
–, normal size 118
–, correlation with sperm count 126, 127
–, echogram vs. clinical size 127
Testosterone 9, 34, 37–40, 66, 80, 133
–, seminal plasma 39, 40
– therapy 74, 137
Thyroid function tests 133
Thyrotoxicosis 95
Tomogram 133
Transvaginal immunization 61
Trichomonas vaginalis 48, 55
Trisomy 8, 84
Tyson's glands 9

Urethral glands 9
Urethritis 53
Urethrocytography 80

Urography 80
Urogenital infections 135
– sinus 9

Valsalva manoeuver 105
Varicocele 105, 112, 121, 124
Varicocelectomy 112
Vas deferens 8, 9, 62
Vas re-anastomosis 113
Vasectomy 113
Vasography 108, 109, 133
Vasopressin 8
Venography 112

X-linked dominant disorders 91
– recessive disorders 91
XX male 81
XXXXY syndrome 74–76
XXXY syndrome 76, 77
XX YY males 78
XY/XO mosaicism 78–80
XYY syndrome 77, 78

Y chromosome, deletion 82
– –, dicentric 82, 83